MW01250598

Hazardous Materials Transportation Risk Analysis

Hazardous Materials Transportation Risk Analysis

Quantitative Approaches for Truck and Train

William R. Rhyne

VAN NOSTRAND REINHOLD
An International Thomson Publishing Company

New York • London • Bonn • Boston • Detroit • Madrid • Melbourne • Mexico City
Paris • Singapore • Tokyo • Toronto • Albany NY • Belmont CA • Cincinnati OH

Library of Congress Catalog Card Number 93-38050
ISBN 0-442-01413-9

I(T)P Van Nostrand Reinhold is an International Thomson Publishing company. ITP logo is a trademark under license.

Printed in the United States of America

Van Nostrand Reinhold
115 Fifth Avenue
New York, NY 10003

International Thomson Publishing
Berkshire House, 168-173
High Holborn, London WC1V 7AA
England

Thomas Nelson Australia
102 Dodds Street
South Melbourne 3205
Victoria, Australia

Nelson Canada
1120 Birchmount Road
Scarborough, Ontario
M1K 5G4, Canada

International Thomson Publishing GmbH
Königswinterer Str. 418
53227 Bonn
Germany

International Thomson Publishing Asia
221 Henderson Building
#05-10 Henderson Building
Singapore 0315

International Thomson Publishing Japan
Hirakawa-cho Kyowa Building, 3F
2-2-1 Hirakawa-cho, Chiyoda-ku
Tokyo 102
Japan

BBR 16 15 14 13 12 11 10 9 8 7 6 5 4 3 2 1

Library of Congress Cataloging in Publication Data

Rhyne, W. R.
Hazardous materials transportation risk analysis: quantitative approaches for truck and train / William R. Rhyne.
p. cm.
Includes bibliographical references and index.
ISBN 0-442-01413-9
1. Hazardous substances—Transportation. 2. Hazardous substances—Risk assessment. I. Title.
T55.3.H3R49 1994
604.7—dc20 93-38050
CIP

To: Bert
Billy
Mike
Maxie
Rut

Contents

Preface

This book is for the transportation professional who wants to know what a transportation quantitative risk analysis is, what it can do, how to communicate risk study objectives to an experienced risk analyst either inside or outside of his/her company, and potentially how to do a reasonably detailed calculation. This user needs to know the definitions of basic terms, the generic quantitative risk analysis procedure and how it is applied to transportation, sources of data specific to transportation quantitative risk analysis, and the methodology for frequency and consequence analysis. The transportation professional needs to understand the uncertainties in the data and methodology, and, most important, how to use the results to reduce risk.

This book also is intended for the quantitative risk analysis professional who wants to know about the special features of a transportation quantitative risk analysis. This user is familiar with quantitative risk analysis concepts, has access to sophisticated consequence methodology, and is knowledgeable about detailed uncertainty and sensitivity analysis techniques.

Neither of these two very different users of this book needs for an exhaustive treatment of consequence analysis to be presented here. Most facets of consequence analysis are not unique to transportation but rather are part of the traditional analysis for stationary chemical plants, and other books and periodicals cover this topic in detail. (The estimation of environmental effects of accidents is considered a subset of consequence analysis.) Similarly, detailed statistical analysis concepts are not presented here.

My colleagues at H&R Technical Associates, Inc. made invaluable contributions both before and during the preparation of this book. I particularly want to thank Denise Crenshaw-Smith, Kathy Goldstein, Ann Hansen, Roy Hardwick, David Johnson, Randy Kirchner, Paul McCluer, Elizabeth Nash, Ken Newman, Randy Robinette, Phil Walsh, and Teresa Wells.

Hazardous Materials Transportation Risk Analysis

1

Quantitative Risk Analysis Concepts

This chapter first provides a set of definitions that establish a standard set of nomenclature. The beginning risk analyst will want to study these definitions, whereas experienced risk analysts probably will want only to ensure that their a priori definitions are consistent with those presented here. Following sections discuss the quantitative risk analysis procedure, some similarities and differences between traditional quantitative risk analysis for stationary chemical plants and for transportation, some limitations to general quantitative risk analysis, and the goals and strategy of this book.

1.1 DEFINITIONS

The terms "hazard" and "risk" are synonymous in everyday usage but are quite different in technical language. A *hazard* is the inherent characteristic of a material, condition, or activity that has the potential to cause harm to people, property, or the environment. A tank pressurized with air has the potential to cause harm to people from fragments should the tank fail. An unpressurized tank filled with a toxic material has the potential to cause harm because of the quantity of toxic material that could be released.

Risk is the combination of the likelihood and the consequence of a specified hazard being realized. *Likelihood* can be expressed as either a frequency or a probability, but in this book it generally will be considered a frequency. *Frequency* is the rate at which events occur and may be expressed as events/year, accidents/mile, and so on. The frequency component of risk often consists of the basic frequency multiplied by several conditional probability terms. *Probability* is a number between zero and one that expresses a degree of belief concerning the possible occurrence of an event. In this book, probability usually refers to a conditional probability. A *conditional probability* is a probability for an event that has been preceded by another specified event. *Consequence* is the direct effect, usually undesirable, of the accident or incident. Consequences usually are measured in health effects but may be expressed as cost of property loss or the amount of hazardous material released.

Risk often is defined as frequency *times* consequence. This expression is a subset of the risk definition presented earlier. When frequency (or probability) is multiplied by consequence, an accident that is expected to cause one fatality and occur 10 times a year has the same mathematical risk as an accident that is expected to cause 1000 fatalities and occur once every 100 years. As can be seen, a great deal of information may be lost when risk is expressed as the product of frequency and consequence.

A *quantitative risk analysis* incorporates numerical estimates of the frequency and the consequences in a sophisticated but approximate manner. In practice, few decisions require quantification at equal levels of sophistication of both frequency and consequence. An *absolute* or *complete quantitative risk analysis* incorporates quantitative estimates of both frequency and consequence, and the consequences are expressed to the full extent (i.e., to health effects). The advantage of an absolute risk analysis is that it could be used to show that a process is safe; that is the calculated risk meets some acceptable risk criteria. One disadvantage of absolute estimates is that certainty about the accuracy of the results is impossible. Also, there are no standard risk acceptability criteria, and the numerical estimates are difficult for nonexperts to interpret (Arendt et al. 1989). A *relative risk analysis* means that a risk is evaluated in comparison with another risk; the comparison can be qualitative or quantitative. Relative risk estimates have the advantage over absolute estimates of being much less likely to be misinterpreted, and the accuracy of their results is easier to defend. The results can be used to rank risk reduction measures. The primary disadvantage of relative estimates is that the results show only that one option is less risky than another, but do not indicate whether one, both, or neither option is acceptable (Arendt et al. 1989).

Risk assessment and risk analysis usually are used interchangeably. The author prefers to define *risk analysis* as the computation of risks and *risk assessment* as the determination of risk acceptability, perhaps by comparison of results with other risks. Taking action to reduce risks is *risk management.*

An *initiating event* is the first in a sequence of events that may lead to an undesirable consequence. An example is the failure of truck brakes to function. If an accident results, *contributing factors* could be excessive speed, poor visibility, and so on. In transportation risk analysis, the initiating event usually is considered a *reportable accident,* an accident that is of sufficient severity to require notification of regulatory bodies. The minimum reportable accident is defined by state or federal regulatory bodies on the basis of the level of property damage or whether an injury or death occurs. An accident usually involves en route vehicles but could also consist of an event such as a vehicle fire while the vehicle is stopped (e.g., overnight).

A reportable accident does not necessarily involve the release of the hazardous material being transported. The U.S. Department of Transportation (DOT) defines an *incident* as a release occurring during loading or unloading, while the vehicle is en route, or when it is in temporary storage related to transportation. Thus, an incident may be unrelated to an accident or may arise from an accident. An accident may or may not result in an incident. In this book we are primarily interested in incidents caused by an accident although sometimes we will consider incidents that occur en route but are not considered accidents. We will refer to the former as accidents that cause a release. An example of the latter is a truck being driven down the highway with a leaking valve. We will refer to these incidents as *non-accident releases.*

1.2 THE QUANTITATIVE RISK ANALYSIS (QRA) PROCEDURE

The QRA procedure is basically the same for transportation as for process system risk analysis. Five basic steps are defined, each of which consists of several substeps as shown in Fig. 1-1. More elaborate versions of Fig. 1-1 are common (CCPS 1989), which emphasize the potential for identifying changes at all levels of analysis to reduce risk.

1.2.1 Preliminary Hazards Analysis

The first activity in a QRA frequently is called a preliminary hazards analysis (PHA). The first step is to define the objectives, scope, and other bounds of the analysis. The first step is equally complex for transportation and for process risk analysis, but the remaining steps in the PHA are much simpler for transportation applications.

1.2.1.1 Define Analysis Objectives and Scope

Several questions can be asked to help define the analysis. The first is "Why is this analysis being done?"; that is, "What decision(s) is to be made by using the results of this analysis?" Possible objectives include the following:

- To estimate bounding or "worst case" consequence with or without a frequency estimate. Any frequency estimate is likely to be qualitative.

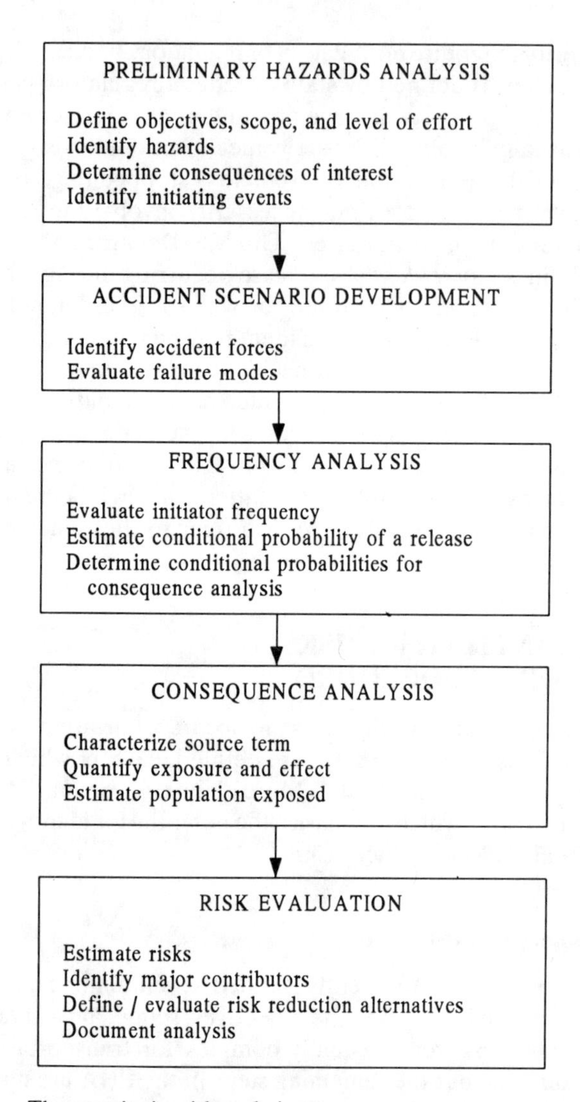

FIGURE 1-1. The quantitative risk analysis process.

- To assist with emergency planning. This objective is essentially the same as the preceding one, and the analysis involves bounding consequence analysis with qualitative frequency analysis.
- To meet regulatory requirements. These analyses tend to be of a bounding level with quantitative frequency analyses to avoid unreasonable focus on "worst case" events.

- To estimate overall corporate risk. Consequences tend to be limited to bounding and other severe effects; however, numerical frequency estimates may be desirable to put consequences in perspective. The level of effort may be identical to that in the preceding objective.
- To evaluate a range of risk reduction options. This objective implies that risks have been bounded with a prior analysis or are being bounded in this analysis with some a priori focus on likely risk reduction options. Risk reduction options at a broad level imply that subsequent, narrowly defined analysis will likely follow. The effort may include one or more transport modes with one or more container types.
- To evaluate a narrow set of risk reduction options. Most likely, prior analyses have indicated major risks, and management has made some selection of the most cost-effective options. The analysis objective is to pinpoint the extent of risk reduction. Accident scenarios not relevant to the options under study may be eliminated entirely or at least restricted. Frequency estimates usually are fairly sophisticated. Consequence analyses are carried only far enough to put all options on a common basis.

The level of analysis seems to have three dimensions: (1) the level of the scenarios to be used—bounding, intermediate, or very detailed; (2) the desired level of frequency quantification—qualitative, detailed quantitative, or somewhere in between; and (3) the desired extent and level of consequence quantification—qualitative, detailed quantitative, or somewhere in between.

A very important question is "How much calendar time and what other resources are available for the analysis?" If plenty of time and resources are available, then the analyst will likely aim at a detailed analysis, so that some contingency is built in to ensure that all of management's questions will be answered. In the great majority of situations, calendar time, resources, or both are in short supply. Thus, the analyst must make trade-offs between cost and/or time and analysis sophistication.

Another consideration that affects the scope of the analysis is the audience to whom the results are to be presented. Are they familiar with the processes being analyzed and/or the risk analysis methods, or must extra effort be planned for analysis description and results presentation?

1.2.1.2 Identify Hazards

For en route analyses, the hazard identification is straightforward: it is the hazardous material being transported. In some cases, hazards may be associated with one alternative under consideration but not associated with another alternative; for example, for some munition types a propellant might be included in one option but only explosives in the other. In a practical analysis, loading and unloading and other facets of the use of the material being transported may be in the project scope. In such a case, hazards can arise from nearby materials or processes that may affect the material being loaded or unloaded.

1.2.1.3 Determine Consequences of Interest

The analysis objectives need to be considered in selecting the consequences to be evaluated. For example, are nonaccident releases important to the objectives? If the analysis purpose is to evaluate head shield puncture resistance, events that lead to releases caused by thermal forces, for example, can be neglected.

1.2.1.4 Identify Initiating Events

The initiating event is the reportable accident or the nonaccident release, if applicable. In some special cases, a specific type of accident (e.g., rollover) may be the initiating event.

1.2.2 Accident Scenario Development

The accident scenario considers the accident initiator, forces that arise from the accident, and how the container reacts to the forces. The two primary techniques for developing transportation accident scenarios are fault trees and event trees. Accident scenario development is presented in Chapter 4, where it is shown that omission of the event tree analysis can cause certain fire scenarios to be overlooked.

1.2.3 Frequency Analysis

Three main approaches are used for transportation frequency analysis. The emphasis in this book is on using cumulative probability distributions of force magnitude along with engineering models to determine failure thresholds for individual accident scenarios. An alternative is to use historical failure data for the specific container(s) typically used for the commodity class (or for the container class) for individual accident scenarios or classes of accident scenarios. The least sophisticated approach is to use a simple historical failure rate for all accidents lumped together for a class of commodities or containers. Databases for each approach are described in Chapter 3.

The frequency analysis consists of determining not only the frequency of the accident initiator (the truck or train accident), but also other factors directly affecting the release scenario (e.g., the probability of a fire), given an accident. These probabilities are presented in Chapter 3. Other conditional probabilities will arise in Chapter 6 from the consequence calculation (e.g., the probability that a certain set of meteorological parameters occurs).

1.2.4 Consequence Analysis

Consequence analysis tools for transportation accident scenarios are the same as those used for stationary process plant analysis. Many different physical processes must be evaluated, and it is rare for one person to be an expert in mechanical failure analysis, atmospheric dispersion, explosion, health effects of exposure, and so on. This part of the risk analysis usually is a multidisciplinary effort (Freeman 1989). An overview of consequence assessment is contained in Chapter 6.

1.2.5 Risk Evaluation and Presentation

Risk evaluation and presentation methods are the same for transportation and process risk analyses. A very important component of this step is evaluating potential risk reduction alternatives. Ideas for risk reduction can result from any of the previous steps. For example, in computing frequencies of accident initiators, obvious "fixes" may occur to the analyst. Regardless of the decision the risk analysis is to support, a primary reason for performing the risk analysis is to generate cost-effective risk reduction suggestions. An overview of risk presentation approaches is contained in Chapter 7.

1.3 COMPARISON WITH CHEMICAL PROCESS RISK ANALYSIS

Many of the similarities of both transportation and stationary chemical process risk analysis are presented in the preceding paragraphs. The fundamental concepts and methodology are the same. Because of the usual singular focus on a single vehicle with a single hazardous cargo, the identification of the hazard, the consequences of interest, and the initiating events are very much simplified for transportation.

More significant differences in the application of the methodology are shown in Fig. 1-2. First and most important, the transport system is not so well defined as a chemical plant. A chemical plant has a specific set of tanks, pipes, pumps, and valves. In a highway transport situation, a truck shares a variety of highway types (e.g., interstate or two-lane roads) with a variety of other vehicles—all driven by different individuals with different attributes, both good and bad. If an accident initiator in a chemical plant is a pump failure, calculation of the system response usually is straightforward. If a truck is in an accident, a number of things can happen. For example, it can

Attribute	Transport	Stationary Plant
System definition	Not well defined	Well defined
Accident scenarios	Few	Many
Population density control	Little	Fences and remote siting
Meteorological conditions	Many sets	One set
Mitigation	Driver and local authorities	Trained plant personnel
Release analysis	Container response to force	System dynamic response

FIGURE 1-2. Differences between transportation and stationary plant risk analyses.

run into a hard (bridge abutment) or soft (motorcycle) object, it can run off a bridge, it can overturn or stay upright, or it can run into a tank truck of flammable material. Not only is the spectrum of transportation accident scenarios very broad, but generally few data exist to help the analyst.

How is this lack of definition handled? Generally, there are fewer accident scenarios in the transport system than in a stationary plant, and they are generally simpler because analysts do not want to, and perhaps cannot, model the enormous complexity of the transport accident environment. In a transport risk analysis, the forces are simplified: perhaps to fire, puncture, impact, and crush, or perhaps to just one force.

The approach used in most process risk analyses is to model the system with fault trees and/or event trees by beginning essentially from scratch because of the vast differences in systems of interest. Then the analyst looks for data to use in the models. Owing to the simplification in accident forces and the similarities in transport containers, it is generally feasible to start a transportation risk analysis with existing fault trees and event trees and then modify them as needed. Containers on wheels have similar accident environments.

The population density around a chemical facility can be controlled by fences and remote siting; so the population can be determined straightforwardly for a risk analysis. The population density around a traffic accident can vary dramatically, for example, from a large city to a rural area. In addition, traffic can build up behind an accident, resulting in a very high population very close to the accident scene. The transportation risk analyst must model a number of situations to account for this difference.

Similarly, each postulated accident site has a set of meteorological parameters including atmospheric stability classes, wind directions, and wind speeds. At a stationary site, the analyst must decide how many parameters from one set to use to provide the needed level of detail. In the case of transportation, the calculational burden is multiplied by the number of postulated accident sites.

When an accident occurs at a processing plant, plant personnel trained in the particularities of the plant and the on-site hazardous materials are available, and they can respond to mitigate the accident consequences. Emergency response personnel in the plant vicinity have an a priori expectation of what will confront them. In a transport accident situation, the driver may or may not be available or may not be trained to support mitigation efforts. The training of emergency response forces for transport accidents will vary widely, as will the time required to respond effectively. The result is to make mitigation modeling more uncertain for a transport accident than it is for a plant accident.

An important difference between transportation and chemical process risk analysis is that in many process plant accident scenarios, a dynamic analysis of a system of pipes, pumps, and so forth is performed to compute flow rates, pressures, and so on. In a transport accident scenario, the response of the container to a force (say impact) is modeled first. Engineering models for container analysis are presented in Chapter 5. Given a rupture in the transport container, a dynamic analysis is used to calculate a release rate.

1.4 LIMITATION OF QUANTITATIVE RISK ANALYSES

An important limitation of risk analysis is that it can provide no guarantee that all accident initiators or accident scenarios have been identified. Whitaker (1991), an experienced risk analyst, believes that the record of predicting failure rates is "not good." Arendt et al. (1989) believe that it is reasonable to expect that trained and experienced analysts will use systematic approaches and will identify the significant risk contributors.

The analyst uses models to describe accident sequences that are mathematically rigorous and potentially sophisticated. However, it is impossible to identify all the factors that can contribute to an accident. Similarly, consequence models are mathematical approximations to, at best, limited experimental data. The models may be validated with one or more experiments, but it cannot be assumed that a model is appropriate for all situations (Arendt et al. 1989).

The accuracy of absolute risk results depends on whether all the significant risk contributors have been included, the realism of the models, and the uncertainty associated with the input data. Uncertainties can be small if the dominant risk contributors can be estimated from adequate historical data, but can be one to two orders of magnitude (or more) if the dominant risk contributors are rare events (Arendt et al. 1989).

The results of a quantitative risk analysis are difficult for independent experts to reproduce. Very different but defendable assumptions often can cause several orders of magnitude difference in the results, even when the "problem" is well defined. For a recent international conference on the risks of transporting dangerous goods, a benchmark problem was specified and seven risk estimation groups contributed risk estimates that varied over several orders of magnitude (Saccomanno et al. 1993).

Despite these limitations, a quantitative risk analysis can provide a rational basis for making decisions regarding transporting hazardous materials safely. The purpose of this book is to guide the experienced process risk analyst and the transportation professional through the transportation quantitative risk analysis procedure. The limitations will not disappear, but they can be managed to minimize their effect.

1.5 GOALS AND STRATEGY OF THIS BOOK

This book is intended for the transportation professional who is a risk analysis novice and who wants to know what questions quantitative risk analysis can answer. It is assumed that should sophisticated calculations of either the frequency or the consequence component be desired, then the transportation professional will either seek the help of a professional risk analyst or will educate him- or herself in advanced risk analysis frequency and consequence calculation methodology elsewhere; for example, the Center for Chemical Process Safety (CCPS) of the American Institute of Chemical Engineers has published several excellent books on the subject (CCPS 1987, 1989). Chapters 1 (on quantitative risk analysis concepts), 6 (on consequence analysis), and 7 (on presentation concepts) of this book are largely generic and are presented to acquaint the transportation professional with general quantitative risk analysis concepts and methods.

The first seven chapters constitute a general discussion of quantitative transportation risk analysis. The goals of this discussion are to present the available methodologies for analyzing transportation risk, to permit judicious selection of a solution procedure for a particular "problem," to recognize the level of detail needed to provide the desired result, and to recognize the assumptions and limitations of commonly used approaches. A detailed numerical example is presented in Chapter 8 to allow the reader to better understand the information in the first seven chapters. The example given is for bulk shipment of chlorine by truck and train. The features of QRA important to transportation are summarized in Chapter 9.

References

Arendt, J. S., D. K. Lorenzo, and A. F. Lusby. 1989. *Evaluating Process Safety in the Chemical Industry—A Manager's Guide to Quantitative Risk Assessment.* Washington, D.C.: Chemical Manufacturers Association.

CCPS (Center for Chemical Process Safety). 1987. *Guidelines for Use of Vapor Cloud Dispersion Models.* New York: American Institute of Chemical Engineers.

CCPS (Center for Chemical Process Safety). 1989. *Guidelines for Chemical Process Quantitative Risk Analysis.* New York: American Institute of Chemical Engineers.

Freeman, R. A. November 1989. What should you worry about when doing a risk assessment? *Chemical Engineering Progress* 85(11):29-34.

Saccomanno, F. F., M. Yu, and J. H. Shortreed. 1993. Risk uncertainty in the transport of hazardous materials. In *Transportation Research Record 1383,* pp. 58-66. Washington, D.C.: Transportation Research Board, National Research Council.

Whitaker, John. 1991. A reappraisal of quantitative risk analysis. *Engineering Management Journal* 3(3):13-16.

2

Transportation Quantitative Risk Analysis

This chapter first presents several reasons for performing a transportation quantitative risk analysis, including the effect of public perception. Following sections give some historical background on transportation risk analysis and describe transportation risk analysis methodologies. The mathematical formulation is given in Section 2.5, and the reduced form used for indicator analysis, particularly in routing, is described in Section 2.6. The chapter concludes with a discussion of uncertainty in the analysis and its effect on the choice of analysis complexity.

2.1 REASONS FOR A TRANSPORTATION QUANTITATIVE RISK ANALYSIS

The DOT has set standards for the containers used for transporting hazardous materials. Compliance with DOT regulations is considered by many to be all that a potential shipper need be concerned with. The inference is that DOT regulatory compliance assures adequate safety. It is true that DOT has set its standards to meet minimum safety levels; however, as described in the following paragraphs, there are several reasons why a shipper still would want to perform a risk analysis. The reasons given below are translated into more specific analysis objectives in Section 1.2.1.1.

2.1.1 Routing Determination

Although transport in compliance with DOT regulations is considered by DOT to be "safe" for any route; some routes are safer than others (FHWA 1992). Regulatory and public interest groups may demand that the best (safest) route be used. In some cases, a choice between multiple modes may be involved in selecting the best route. Selection of the safest route may become a regulatory requirement; some states (Colorado and California) have established routing procedures, and Ohio has passed a law to require routing for trucks carrying hazardous materials. A notice of proposed rulemaking that would require routing throughout the United States has been issued (FHWA 1992). Even if it is not required, from the shipper's viewpoint, prudence may dictate the determination and the use of the safest route. The type of risk analysis typically used for a routing analysis is a relative risk analysis. See Section 2.6 for caveats about performing routing risk analysis using risk indicators as a relative risk measure.

2.1.2 Public Relations

In today's regulatory, public relations, and litigation environments, it is highly desirable for the shipper to be able to talk intelligently about the potential effects of low-frequency accidents. Therefore, some type of analysis should be performed other than the classical "worst case" analysis. Those who oppose the shipper's plans will use the worst case approach. The shipper would be well advised to have performed consequence analyses for a spectrum of unusual accidents and to have at least a qualitative estimate for the frequencies associated with the specific accidents. Armed with good analyses, the shipper is prepared to deflect criticism of proposed actions. As an example of the problems that can occur, liquefied natural gas facilities in California never were built, at least partially because a worst case analysis was not put in perspective by an analysis of the event probability (Kunruether and Linnerooth 1984).

2.1.3 Liability Control

The shipper can evaluate potential financial exposure should an unusual accident occur. Risk analysis provides quantitative estimates of the potential consequences and the associated frequencies of these consequences. If an accident should cause a shipper to end up in litigation, the opposition may attempt to show that meeting only minimum safety levels was not prudent (Bierlein 1991).

2.1.4 Risk Management

The shipper can reduce exposure to unusual accidents. Risk analysis can predict the container changes and/or administrative controls that will result in the greatest risk reduction. Aided by information on the costs of various options, the shipper can make cost-effective decisions. For example, is it better to have increased impact protection or increased fire protection? In this context "better" means the most risk reduction at the least cost.

2.1.5 Summary

Quantitative risk analysis provides a basis for making consistent and defendable decisions. It allows the evaluation of the effectiveness of existing controls and procedures, but, more important, such analysis provides insight on how to cost-effectively reduce the risk. This book addresses all these reasons for conducting transportation quantitative risk analysis. The last reason, cost-effective reduction of risk, is potentially the most important, and it is specifically the reason why this book was written.

2.2 PUBLIC PERCEPTION OF HAZARDOUS MATERIALS TRANSPORTATION

The need to make consistent and defendable decisions arises from potential adverse public and/or regulatory reaction to hazardous materials shipments. Is this concern justified when approximately 500,000 movements of hazardous materials occur each day by land, sea, and air in the United States (FHWA 1992)? Clearly, hazardous materials are transported safely every day; but the public is concerned because even though major accidents occur infrequently, one accident has the potential to cause substantial health effects and property damage.

In some cases, the threat of problems, given that an accident has occurred, can cause major stress even if the situation ultimately is resolved without serious loss. Table 2-1 shows the number of population evacuations caused

TABLE 2-1. Chemical accident evacuations by cause and year

Cause of evacuation	1980	1981	1982	1983	1984	Totals
Train derailment	14	8	13	12	8	55
Train car spill/fire	3	6	5	4	5	23
Truck accident	9	9	6	6	5	35
Truck spill/fire	1	11	4	9	7	32
Chemical plant release	5	10	15	8	5	43
Industrial plant release	3	10	18	23	24	78
Pipeline	2	1	1	0	0	4
Ship accident	2	1	0	0	1	4
Waste site accident	0	1	2	3	1	7
Other	4	5	4	0	1	14
Totals	43	62	68	65	57	295

Source: Sorensen 1987.

by chemical accidents in the United States for each of the years from 1980 through 1984 (Sorensen 1987). Approximately half of these were due to truck or train accidents. Transportation-related evacuations average about 30 per year. Media attention to these events expands the consciousness of the general public, perhaps overly so.

2.3 HISTORICAL BACKGROUND OF TRANSPORTATION RISK ANALYSIS

In 1961, Leimkuhler et al. performed a statistical analysis of the frequency and severity of truck accidents. Regression analyses were used to determine accident frequency as a function of cargo type (e.g., petroleum) and season of the year. Cumulative probability distributions were determined for the net impact velocity of collisions with other trucks and with automobiles, each at different angles of impact, and for overturns and other collisions combined. The distributions of various collision accident types and other accident types were tabulated for four-lane, divided highways; for low-traffic (under 5000 vehicles per day), two-lane roads; for high-traffic, two-lane roads; and for combined intercity roads. The probability of fire occurring was presented by collision type (1% for all collisions combined), for noncollisions (6.5%), and for all accidents combined (1.8%). Lognormal correlations were derived for the cumulative probability distribution of vehicle damage costs for different collision types. Assuming that collisions are inelastic and using conservation of energy and momentum equations, the transport vehicle damage cost was correlated with the mechanical energy of the accident. Probability distributions were determined for the percent of cargo damaged for collision and noncollision accidents. The only data available to Leimkuhler et al. (1961) for cargo damage were for new automobile transport.

TABLE 2-2. Accident severity categories

Severity	Vehicle speed, mph	Fire duration, hr
Minor	0-30	0-½
	30-50	0
Moderate	0-30	½-1
	30-70	<½
Severe	0-50	>1
	30-70	½-1
	>70	0-½
Extra severe	50-70	>1
	>70	½-1
Extreme	>70	>1

Source: USAEC 1972.

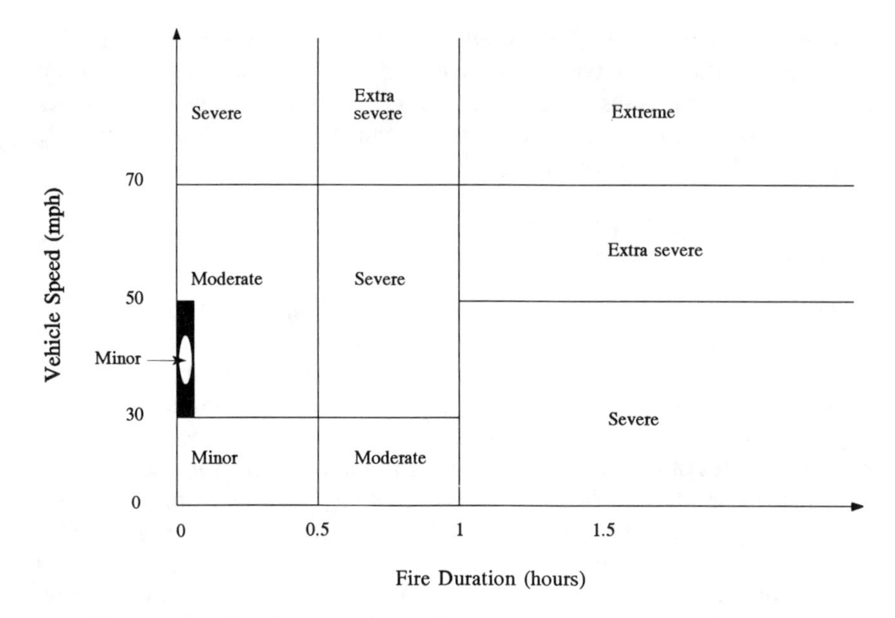

FIGURE 2-1. Graphic display of accident severity categories.

An updated analysis (Leimkuhler 1963) included a correlation of cargo damage with transport vehicle damage for trucks transporting new automobiles and bulk petroleum products. Vehicle damage was correlated to highway type and accident type. Risk reduction investigations included controls on truck speed and rerouting to avoid congested two-lane roads.

In the late 1960s and early 1970s, Bill Brobst, Transportation Branch Chief for the U.S. Atomic Energy Commission, developed the scheme shown in Table 2-2 for categorizing rail and highway accident severity in terms of impact speed and fire duration (USAEC 1972). Figure 2-1 shows the accident severity scheme graphically. Table 2-3 shows Brobst's values for accident frequencies associated with the accident severity subcategories (USAEC 1972), which were based on Leimkuhler's speed distributions. A key feature of Brobst's work is the conditional probability of container damage, given an accident of a particular severity. Brobst (1973) presented such estimates for 55-gal drums (Table 2-4) and for very thick-walled (more than 6 in. thick) containers for highly radioactive materials. A great deal of engineering judgment is incorporated into the values presented in Table 2-4; Brobst (1973) describes the values as based on "gross judgment and the very limited data available." Some analysts still recommend these data for hazardous

TABLE 2-3. Accident frequency for severity categories

Severity category	Vehicle speed, mph	Fire duration, hr	Frequency per Vehicle Mile	
			Rail	Truck
Minor	0-30	<½	6×10^{-9}	6×10^{-9}
	0-30	0	4.7×10^{-7}	4×10^{-7}
	30-50	0	2.6×10^{-7}	9×10^{-7}
Total			7.3×10^{-7}	1.3×10^{-6}
Moderate	0-30	½-1	9.3×10^{-10}	5×10^{-11}
	30-50	<½	3.3×10^{-9}	1×10^{-8}
	50-70	<½	9.9×10^{-10}	5×10^{-9}
	50-70	0	7.5×10^{-8}	3×10^{-7}
Total			7.9×10^{-8}	3×10^{-7}
Severe	0-30	>1	7.0×10^{-11}	5×10^{-12}
	30-50	>1	3.9×10^{-11}	1×10^{-11}
	30-50	½-1	5.1×10^{-10}	1×10^{-10}
	50-70	½-1	1.5×10^{-10}	6×10^{-12}
	>70	<½	1×10^{-11}	1×10^{-10}
	>70	0	8×10^{-10}	8×10^{-9}
Total			1.5×10^{-9}	8×10^{-9}
Extra severe	50-70	>1	1.1×10^{-11}	6×10^{-13}
	>70	½-1	1.6×10^{-12}	2×10^{-13}
Total			1.3×10^{-11}	8×10^{-13}
Extreme	>70	>1	1.2×10^{-13}	2×10^{-14}

Source: USAEC 1972.

TABLE 2-4. Predicted fraction of steel drums damaged vs. accident severity

Damage*	Accident Severity Category				
	Minor	Moderate	Severe	Extra severe	Extreme
None to moderate (no breach)	0.90	0.80	0.55	0.25	0.10
Severe (small breach)	0.07	0.05	0.30	0.25	0.10
Extra severe (medium breach)	0.02	0.03	0.10	0.30	0.30
Extreme (large breach)	0.01	0.02	0.05	0.20	0.50

Source: Brobst 1973.

*Small breach—small amount of gases or liquids released. Medium breach—most gases and liquids and a small fraction of the solid contents released. Large breach—all gases and liquids and a few percent of the solid contents released.

chemical transportation risk analysis (Fenstermacher et al. 1987). An updated accident categorization approach is used for some nuclear material transportation risk analyses (Neuhauser and Kanipe 1992).

An early application of fault tree methodology to transportation risk analysis was made by Garrick et al. (1969). The fault trees were used to define accident scenarios that could cause failure of the container. Many fault tree analyses in the early 1970s were extensions of those performed by Garrick et al. (1969) and used accident frequencies for accident severity categories based on those given in Table 2-3. Hodge and Jarrett (1975), for example, considered container failure mechanisms of impact, puncture, thermal effects, vibration, equipment failure, and operator error. The probability of container failure was estimated by using engineering judgment for each failure mechanism for each accident severity category. Each release could be classified as small, moderate, or large, and a tabular correlation between accident severity category and release class similar to that developed by Brobst was used.

A major advance in quantitative risk analysis capability occurred in 1976 with the publication by researchers at the Sandia National Laboratories (SNL) of probabilistic distributions for the magnitude of mechanical and thermal threats resulting from accident forces (Clarke et al. 1976). Figure 2-2 shows such a distribution for impact forces in train accidents from a second SNL report (Dennis et al. 1978). [The approach of the SNL researchers is similar in some respects to the one used by Leimkuhler et al. (1961) with the major exception being use of engineering models rather than the limited available data for cargo damage.] The accident environments considered to pose the greatest threat to a transport container were impact, crush, puncture, fire, and immersion. Data similar to those presented in Fig. 2-2 were given for transport by aircraft, truck, and train. The container was considered to be roughly the size of a 55-gal drum, and engineering models of the containers in a vehicle, for example, a spring-mass model, were combined with accident statistics to develop accident severity results such as those presented in Fig. 2-2. In 1978, the companion report (Dennis et al. 1978) was published for large containers, that is, those roughly the size of a truck tanker or a train tank car. The probabilistic accident severity distributions in these two reports still are used extensively. (These models and probabilistic results are described in detail in Chapter 3.)

Starting in 1975, researchers at the Pacific Northwest Laboratories (PNL) began issuing reports analyzing the risk of transporting a variety of nuclear materials, the first being for plutonium (McSweeney et al. 1975). The PNL methodology is based on a fault tree analysis to define accident scenarios, the SNL accident severity results, and package failure thresholds determined from engineering analysis. Results are presented as frequency versus conse-

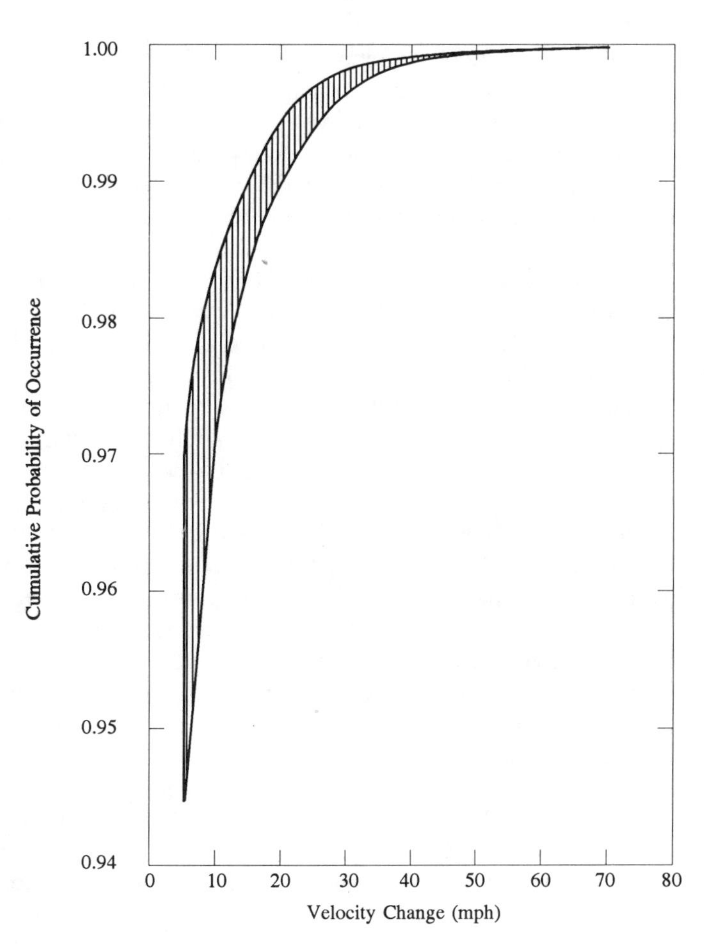

FIGURE 2-2. Cumulative probability distribution of the impact velocity magnitude for large containers in train accidents. Source: Dennis et al. 1978.

quence curves. Researchers at PNL also prepared risk analyses for transporting the following nonradioactive materials: chlorine (Andrews et al. 1980), propane (Geffen et al. 1980), and gasoline (Rhoads et al. 1978). Results of these analyses frequently are cited in the literature.

2.4 QUANTITATIVE RISK ANALYSIS METHODOLOGIES

Risk is defined in Chapter 1 in terms of two parameters: the likelihood (frequency or probability) of occurrence of an accident scenario and the magnitude of the accident consequence. Rowe (1983) characterizes quantitative risk analysis methodologies for transportation in three ways: (1) how they combine the two parameters to arrive at risk; (2) the level of detail; and (3) the methods for obtaining data and modeling parameters.

Rowe gives as an example of combining risk parameters the reduction of the consequence component to simply the occurrence of an accident; thus "risk" is the "frequency of an accident." Another parameter combination example Rowe gives is the use of the expected value of risk: the product of accident scenario frequency and the consequence magnitude. Rowe's other two characterization attributes are of more interest for this book.

For level of detail, Rowe describes the "bottom-up" approach, as beginning at the smallest risk component and progressing upward by combining results to obtain an overall risk. In contrast, other applications start at a high level of data aggregation, that is, a low level of risk component detail. For data sources and modeling, Rowe characterizes the use of fault trees at one end of the scale and average accident rates at the other end. The bottom-up approach is used as an example of a high level of data and modeling complexity. Rowe describes bottom-up as a "fine-grain" estimate made up of a large number of computations, and "top-down" as the aggregation of data to find "cause and effect relationships that can then be used to find overall values" for conditional probabilities. As pointed out by Rowe, there are major difficulties in obtaining data for use of either the top-down or the bottom-up approach, particularly for an absolute risk analysis.

The probabilistic or quantitative risk methodologies presented in this book are best described by the level of aggregation, that is, bottom-up or top-down. All quantitative risk analyses aggregate risk-producing components to some level. Before this characterization of risk methodology is considered, the two parameters, frequency and consequence, must be defined in some detail.

The frequency component of risk can be considered as consisting of three subcomponents: the accident frequency, the conditional probability of the release of contents given that an accident has occurred, and conditional probabilities that arise from the consequence component such as the meteorological conditions. The consequence component of risk can be considered to consist of three subcomponents: the amount of material released, the number of people exposed, and the health effect(s) of the exposure. These six subcomponents frequently are further divided. The procedure for numerically evaluating each subcomponent is dependent on the purpose for which the evaluation is being performed. If container design aspects such as valve protection features or head shields are of interest, then the dependence of the risk parameters on these container design features should be evaluated explicitly. If administrative controls, such as special speed controls, are the likely source of risk reduction, then these controls should be explicitly evaluated in the risk parameters.

Two potential sequences for evaluating quantitative risk are shown in Fig. 2-3. The top sequence begins with the accident frequency (F_1). The conditional probability of contents release is subdivided into the conditional probability for occurrence of a type of accident force (P_2), the conditional probability that a certain magnitude of that accident force occurs (P_3), and the conditional probability that the magnitude is sufficient to cause container failure (P_4). The consequence components consist of the release amount (A), the dispersion of released material that causes exposure to persons and the health effect of exposure (X), and the number of people exposed (N). (Not shown are conditional probabilities associated with the consequence components.) Each block in the figure can be evaluated as a function of the variables indicated; for example, the conditional probabilities of the forces that occur from an accident are a function of the transport mode. (Actually the independent variables are numerous but usually are aggregated into only mode-dependent values.) The lower sequence consists of the accident frequency, an expression of the conditional probability of a release amount, the health effect of exposure, and the number of people exposed. To use the lower sequence, the conditional probability and the amount of release must be developed a priori. Such a correlation is presented in Nayak et al. (1983) for rail, including variation with track class, train speed, and hazardous material class. If computation of the risk of train transport of gasoline is desired, the analyst must decide if the extra computational effort involved in the top sequence is cost-effective compared to evaluating the lower sequence for flammable liquids as a class of hazardous material. The top sequence is presumably more accurate but more costly to compute; the lower sequence is presumably less accurate but, if the frequency/amount correlation is readily available, more easily computed. The top sequence is useful when detailed design features are of interest, and the bottom sequence is useful only for determining risk for, in this case, a broad class of railcar/commodity combinations. For the particular correlation indicated in Fig. 2-3, risk mitigation effects can be determined only for variations in track class, train speed, or hazardous material class. The analyst must make the required decision concerning which calculational sequence to use for the particular problem being evaluated.

Continuing with the gasoline tank car example, although there are data that permit the correlation of frequency and amounts of release for broad classes of materials, there are not sufficient accident data to correlate the frequency of release from a gasoline tank car for various valve parameters.

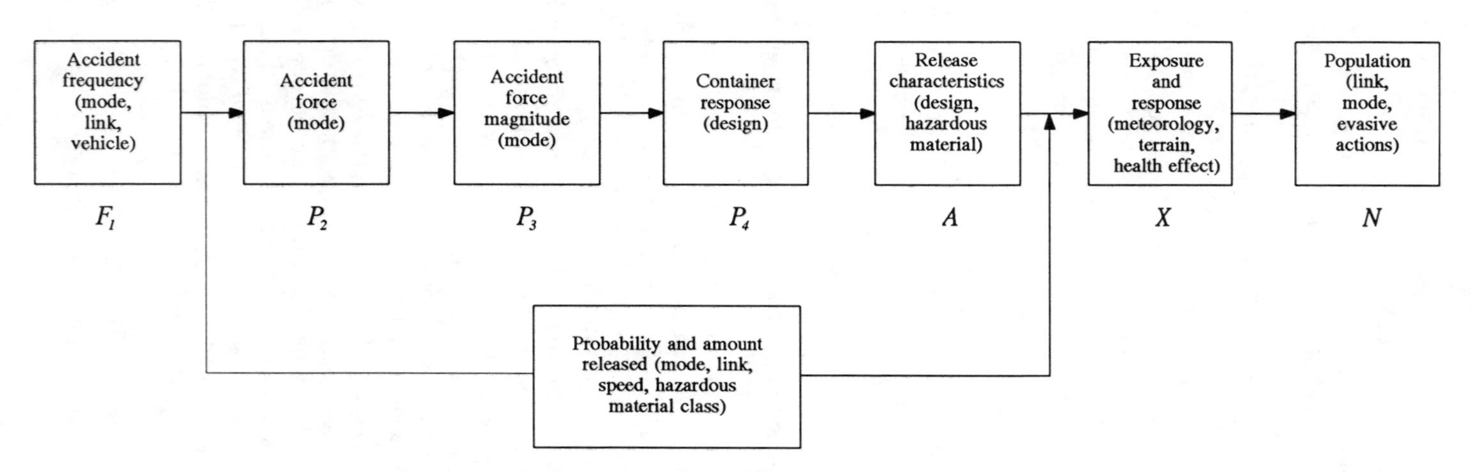

FIGURE 2-3. Two possible sequences for evaluating transportation quantitative risk.

Thus, the risk analyst must have techniques to predict the effect on risk of changes in valve parameters for gasoline tank cars because historical data are not available. This choice of approach illustrates the two broad classes of quantitative transportation risk analysis: predictive and historical. Generally the predictive approach is synonymous with the bottom-up method and the historical with the top-down method.

The choice between the predictive and historical approaches is illustrated in Fig. 2-4. If data are available that apply directly to the situation, they should be used. However, if the data are not applicable, an analyst using the predictive approach would use engineering models and the available data to construct an approximation for the needed parameter from the bottom up. An analyst using the historical approach would broaden the definition of the hazardous material to include other materials of similar behavior, or broaden the definition of the container to include other similar containers, and so on, until enough historical data were included in the definition to permit derivation of a statistically significant value for the needed parameter. That is, the historical analysis proceeds from the top down until enough data are aggregated. This book will focus on the more computationally complex predictive or bottom-up approach to quantitative risk analysis; however, data

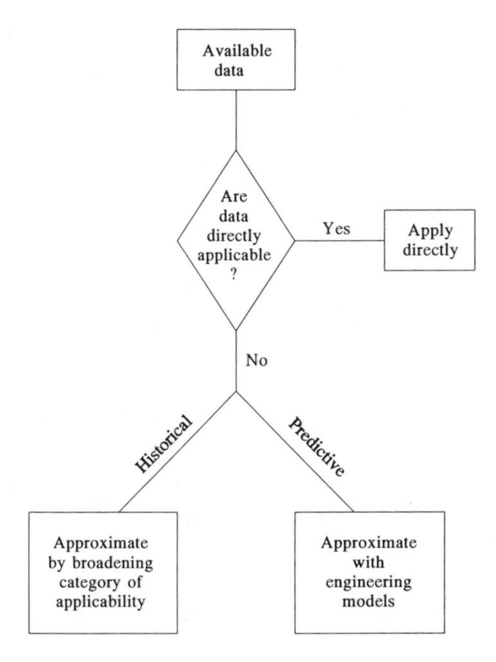

FIGURE 2-4. Comparison of historical and predictive approaches.

sources for both approaches will be addressed in Chapter 3. In practice some parameters are almost always evaluated using historical data, for example, accident rate, so the distinction focuses on how the other parameters are evaluated, particularly the container failure probability.

2.5 MATHEMATICAL FORMULATION

The risk, R_i, for accident scenario i is a function of the scenario frequency, F_i, and the scenario consequence, C_i:

$$R_i = f(F_i, C_i) \tag{2-1}$$

The usual procedure for a quantitative transportation risk analysis is to divide the transport route into segments (also called links) along which the important parameters can be reasonably approximated by a single average value. A detailed expression for risk then can be further defined:

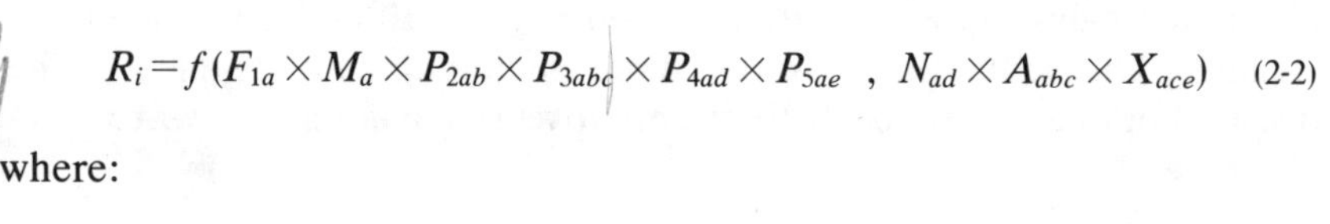

$$R_i = f(F_{1a} \times M_a \times P_{2ab} \times P_{3abc} \times P_{4ad} \times P_{5ae}\ ,\ N_{ad} \times A_{abc} \times X_{ace}) \tag{2-2}$$

where:

- F_{1a} = frequency of an accident per mile in transport link a based primarily on highway (or rail track) type and conditions, vehicle type, and traffic conditions
- M_a = number of miles, or miles per year, in link a
- P_{2ab} = probability that the accident in link a results in accident forces of type b (e.g., mechanical or thermal forces)
- P_{3abc} = probability that release class c occurs, given that the accident force type b occurs in link a, which depends on the force magnitude and the container's capability to resist the force
- P_{4ad} = probability that population distribution class d occurs in link a
- P_{5ae} = probability that meteorological condition e occurs in link a
- N_{ad} = number of persons per unit area in population class d in link a
- A_{abc} = release amount for release class c, given that force type b occurs in link a
- X_{ace} = area that experiences the specified health effect from a unit release of the hazardous material for meteorological condition e for release class c

The overall risk is obtained by summing all scenarios for each link or for the entire route:

$$R = \Sigma R_i \tag{2-3}$$

The extent to which the risk components just given are aggregated before numerical values are assigned depends on the reason why the risk analysis is being performed and the resources (time and personnel) available to the analyst. (See also Sections 1.2.1.1 and 2.7.)

2.6 RISK INDICATORS

Equations 2-2 and 2-3 will produce a quantitative value for absolute (or complete) risk. It is frequently useful to compute relative risk for two or more options using only a few of the parameters from Eq. 2-2 as a surrogate for risk, an approach often used for comparing routing options. Historically, the basic approach has been to use the accident rate per mile times the number of miles in a highway segment as a surrogate for the frequency portion of Eq. 2-2 and the number of people in a 0.5- to 1-mile-wide band along the highway segment as a surrogate for the consequence portion of Eq. 2-2. The product of the two terms is a relative risk indicator, and the route with the lowest indicator has the lowest computed relative risk. The DOT Research and Special Programs Administration (RSPA) guide for routing analysis (RSPA 1989) allows routes to be excluded for physical or legal reasons. Factors that are not easily quantified, such as the presence of schools or hospitals that are not easily evacuated or the presence of a reservoir, can be considered subjectively. A recent notice of proposed rulemaking (FHWA 1992) lists 13 elements that should be considered in establishing highway routing requirements:

1. Population density.
2. Type of highway.
3. Type and quantities of hazardous materials.
4. Emergency response capabilities.
5. Results of consultations with affected persons.
6. Exposure and other risk factors (such as distance to hospitals).
7. Terrain considerations.
8. Continuity of routes.
9. Alternate route.
10. Effects on commerce.
11. Delays in transportation.
12. Climatic conditions.
13. Congestion.

Abkowitz et al. (1992) have evaluated the use of five criteria for routing analysis: (1) minimize shipping distance, (2) minimize travel time, (3) minimize release-causing accident likelihood, (4) minimize population exposure, and (5) minimize the product of accident likelihood and population. The first two criteria minimize economic cost, and the latter three maximize safety. Abkowitz et al. (1992) found that routes that minimize risk may be so circuitous that they can be economically unfeasible, or at least impractical. The recommendation is that a routing analysis consider combinations of factors and use different weighting factors to evaluate trade-offs. For example, a route selected by using travel time minimization weighted by 0.75 and risk minimization weighted by 0.25 resulted in only a 3% increase in travel time over the route determined to produce minimum travel time, yet it reduced the risk associated with the minimum travel time by 70%.

The risk indicator approach can be a useful routing tool. The use of the reduced form of Eq. 2-2 involves some implicit assumptions that the user should be aware of. The first assumption is that risk is the product of frequency and consequence; thus, the lowest product of the terms of Eq. 2-2 defines the option with the lowest risk. For simplicity of presentation, the following additional assumptions will be used: only one release class ($c = 1$), one population distribution along each link type ($d = 1$), and one meteorological condition ($e = 1$). Thus, $P_4 = P_5 = 1$. If the analyst wishes to compare options x and y, then the question is whether R^x is less than, greater than, or equal to R^y. Using Eq. 2-2, the question can be reformulated as follows:

$$\text{Compare } F_1^x M^x P_2^x P_3^x A^x X^x N^x \text{ and } F_1^y M^y P_2^y P_3^y A^y X^y N^y \tag{2-4}$$

If some terms are the same for both options (e.g., if $P_2^x = P_2^y$, $P_3^x = P_3^y$, $A^x = A^y$, and $X^x = X^y$), then Eq. 2-4 is simplified to the following expression:

$$\text{Compare } F_1^x M^x N^x \text{ and } F_1^y M^y N^y \tag{2-5}$$

The remainder of this section will address the explicit and implicit assumptions involved in simplified forms such as Eq. 2-5.

A routing study for a single transport mode could be based on minimizing the product $F_1 MN$ (accident rate, number of miles, and exposed population). Use of this simple risk indicator for risk minimization includes some important assumptions. Clearly, the approach would be valid only if the same container were used on all potential routes; thus, the container failure thresholds and the container response to the accident force types would be the same. On this basis, the P_3 parameter (the probability that the container will fail from the accident force) is neglected; however, this omission also implies that the force magnitude is the same for all potential routes. For most routing decisions, it probably is not practical to try to include route-dependent variations in the frequency and the magnitude of mechanical threats (e.g., bridge abutments or rock outcrops). On the other hand, the model usually used for estimating the magnitude of the threat from fire (Dennis et al. 1978) explicitly includes a factor for the effectiveness of the response of a local fire department. The omission of the P_3 factor, therefore, invokes the assumption that all routes have equal fire-fighting response capability.

The P_2 parameter represents the distribution between the various accident threat types (e.g., impact, puncture, crush, fire, and immersion). Data generally are not readily available to make a distinction between the relative distribution of accident force types as a function of road type or track class; therefore, omission of the P_2 term may be a practical necessity. The P_2 and P_3 terms together represent the probability of container failure, given an accident. Some data have been presented on the probability of a hazardous material release, given an accident, as a function of highway class and as a function of urban or rural demography (Harwood et al. 1990). Presumably, such data account for more than fire-fighting response variation; so the use of F_1MN as a risk measure ignores any potential variation in release probability, given an accident, as a function of road class, population density, or both.

The computed release amount, given a release, is a strong function of the level of detail in the engineering analysis of the response of the container to the accident forces and a weak function (if any) of the transport link type. Thus, the omission of the A parameter for routing analyses involving only one transport mode is a practical approach.

The X parameter represents the effect of the released material on the population in the vicinity of the accident. For hazardous materials whose release affects the surrounding area by a downwind plume, omission of this term implies that meteorological parameters are the same for all routes being considered.

Generally, simple risk indicators used for routing analyses can be considered as reduced forms of the complete quantitative risk equation. Simple risk indicators do not provide as much information regarding the safety of hazardous material transport as do complete quantitative risk analyses. However, simple risk indicators can be useful in some decisions involving two or more alternatives. The appropriate calculational approach is a function of the decision to be made, practical constraints such as data availability, and budgetary resources available. Before one decides to use a simple risk indicator, the complete quantitative risk formulation should be the starting point, and terms should be eliminated only after careful consideration of the implicit and explicit assumptions involved in their omission.

2.7 UNCERTAINTY AND THE CHOICE OF ANALYSIS COMPLEXITY

The choice of analysis approach and level of complexity depends primarily on the decision to be made, that is, the question that the analyst seeks to answer. In some situations, a constraint is that not enough calendar time, financial resources, and/or technical resources are available to provide the answer in the desired detail. If a very rapid analysis is needed, perhaps only a few variables can be included as surrogates for a more complete analysis. An alternatives analysis should focus on the elements that will help distinguish between the proposed alternatives; precision among the results for the alternatives is more important than accuracy on an absolute scale.

The data and the analysis models currently available may introduce large uncertainty, up to several orders of magnitude. Recent analyses of a sample problem by international teams underscore the large differences that arise from the use of different models and assumptions (Saccomanno et al. 1993). The analyst must consider whether incorporating the full complexity of relationships in the analysis is worthwhile; is it practical when uncertainties are not well understood?

The analysis can be detailed or approximate. Usually an approximate model will be sufficient, but the recommended approach is to favor a more detailed methodology, particularly for the variables of interest in decision making. The analyst can more easily collapse a detailed model than expand a simple model. Furthermore, if a more precise analysis is warranted later, the basis will have been established. There will be uncertainty, and certain steps in the analysis will contribute to the uncertainty disproportionately. The analysis procedure (Figs. 1-1 and 2-3) is a chain that is subject to the "weak link" argument. The analyst should provide detailed, cost-effective results, but all assumptions and bases must be clearly delineated.

References

Note: The reports of U.S. government agencies, their laboratories, and contractors cited here are available from the National Technical Information Service, Springfield, Virginia 22161, USA.

Abkowitz, M., et al. 1992. Selecting criteria for designating hazardous materials highway routes. In *Transportation Research Record 1333,* pp. 30–35. Washington, D.C.: Transportation Research Board, National Research Council.

Andrews, W. B., et al. March 1980. *An Assessment of the Risk of Transporting Liquid Chlorine by Rail.* PNL-3376. Pacific Northwest Laboratory.

Bierlein, L. W. November 1991. Hazmat packaging: don't cut corners. *Traffic Management* 30(11):75.

Brobst, W. A. May 1973. Transportation accidents: how probable? *Nuclear News* 16(5):48–54.

Clarke, R. K., et al. July 1976. *Severities of Transportation Accidents.* SLA-74-0001. Sandia National Laboratories.

Dennis, A. W., et al. May 1978. *Severities of Transportation Accidents Involving Large Packages.* SAND77-0001. Sandia National Laboratories.

Fenstermacher, J. E., et al. November 1987. *A Manual for Performing Transportation Risk Assessment.* PLG-0588. Washington, D.C.: Chemical Manufacturers Association.

FHWA (Federal Highway Administration). August 31, 1992. Transportation of hazardous materials: highway routing. *Federal Register* 57(169):39522-39533.

Garrick, B. J., et al. 1969. *A Risk Model for the Transport of Hazardous Materials.* HN-204 (available from the Defense Documentation Center as AD-860,120 on a limited distribution basis). Washington, D.C.: Department of the Army.

Geffen, C. A., et al. March 1980. *An Assessment of the Risk of Transporting Propane by Truck and Train.* PNL-3308. Pacific Northwest Laboratory.

Harwood, D. W., J. G. Viner, and E. R. Russell. 1990. Truck rate model for hazardous materials routing. In *Transportation Research Record 1264,* pp. 12-23. Washington, D.C.: Transportation Research Board, National Research Council.

Hodge, C. V., and A. A. Jarrett. April 1975. *Transportation Accident Risks in the Nuclear Power Industry 1975-2000.* EPA-520/3-75-023. Environmental Protection Agency.

Kunreuther, H., and J. Linnerooth. June 1984. Low probability accidents. *Risk Analysis* 4(2):143-152.

Leimkuhler, F. F. 1963. *Trucking of Radioactive Materials: Safety vs. Economy in Highway Transport.* Baltimore, Maryland: The Johns Hopkins Press.

Leimkuhler, F. F., M. J. Karsen, and J. T. Thompson. 1961. *Statistical Analysis of the Frequency and Severity of Accidents to Potential Highway Carriers of Highly Radioactive Materials.* NYO-9771. U.S. Atomic Energy Commission.

McSweeney, T., et al. August 1975. *An Assessment of the Risk of Transporting Plutonium Oxide and Liquid Plutonium Nitrate by Truck.* BNWL-1846. Pacific Northwest Laboratory.

Nayak, P. R., et al. November 1983. *Event Probabilities and Impact Zones for Hazardous Materials Accidents on Railroads.* DOT/FRA/ORD-83/20 (available as PB85-149854). Federal Railroad Administration.

Neuhauser, K. S., and F. L. Kanipe. 1992. *RADTRAN 4: A Computer Code to Analyze Transportation of Radioactive Materials, Volume 3, Users Guide.* SAND89-2370 (TTC-0943). Sandia National Laboratories.

Rhoads, R. E., et al. November 1978. *An Assessment of the Risk of Transporting Gasoline by Truck.* PNL-2133. Pacific Northwest Laboratory.

Rowe, W. D. November 1983. *Risk Assessment Processes for Hazardous Materials Transportation.* Transportation Research Board National Cooperative Highway Research Program Synthesis of Highway Practice Report 103. Washington, D.C.: Transportation Research Board, National Research Council.

RSPA (Research and Special Programs Administration). July 1989. *Guidelines for Applying Criteria to Designate Routes for Transporting Hazardous Materials.* DOT/RSPA/OHMT-89-02. U.S. Department of Transportation.

Saccomanno, F. F., M. Yu, and J. H. Shortreed. 1993. Risk uncertainty in the transport of hazardous materials. In *Transportation Research Record 1383,* pp. 58-66. Washington, D.C.: Transportation Research Board, National Research Council.

Sorensen, J. H. 1987. Evacuations due to off-site releases from chemical accidents: experience from 1980 to 1984. *Journal of Hazardous Materials* 14(2):247-257.

USAEC (U.S. Atomic Energy Commission). December 1972. *Environmental Survey of Transportation of Radioactive Materials to and from Nuclear Power Plants.* WASH-1238.

3

Databases

This chapter addresses data sources for each component of the transportation risk expression from Chapter 2 that is unique to transportation risk analysis. Most frequency-related parameters are included: accident rate (Section 3.1), accident force type and accident force magnitude (Section 3.2), and container failure probability (Section 3.3). Only one consequence-related parameter is included: release rate and/or amount (Section 3.4).

3.1 ACCIDENT RATE OR FREQUENCY

The accident rate is the number of accidents per unit of highway or track length. It is expressed as:

$$\text{accident rate (accidents/mile)} = \frac{\text{Number of vehicle accidents}}{\text{number of vehicle miles}} \tag{3-1}$$

The accident rate is computed by dividing the number of accidents that have occurred by the corresponding exposure measure of opportunities for an accident to occur, that is, the number of vehicle miles. Data are available for both the number of accidents and the number of vehicle miles; however, collection procedures and quality vary. The Office of Technology Assessment (OTA) concluded in 1986 that data collection and analysis capability needs improvement for all modes: highway, rail, waterway, and air. Highway data were considered the worst (OTA 1986). The present capability to compute this parameter as a function of highway or track characteristics is presented in Sections 3.1.1 and 3.1.2.

3.1.1 Truck

Data needed for the computation of truck accident rates are collected by federal agencies, such as the National Highway Traffic Safety Administration (NHTSA) and the Federal Highway Administration (FHWA), all of the states, industrial organizations, and research groups, such as the University of Michigan Transportation Research Institute (UMTRI). The most commonly cited sources of truck accident data and their strengths and weaknesses are presented in Table 3-1 (TRB 1990). Until March 4, 1993, the definition of a reportable accident in Title 49 of The Code of Federal Regulations, Part 390 was the death of a person, bodily injury to a person that requires immediate medical treatment away from the accident scene, or property damage of $4,400 or more. March 4, 1993, was the effective date for the FHWA replacement of the property damage criterion by the criterion of disabling damage, which is defined as damage that precludes departure in the usual manner from the accident scene after simple repairs (58 FR 6726). The Motor Carrier Accident Report 50-T listed in Table 3-1 also was discontinued on March 4, 1993.

The annual number of fatal truck crashes is recorded in a reasonably reliable way, but the nonfatal truck crashes are not well recorded. For example, of the existing databases that include both fatal and nonfatal crashes, the reported number of crashes involving combination trucks (trucks or tractors pulling one or more trailers) varied from 33,000 to 1,200,000 for 1985, the last year for which data were available from all three databases (TRB 1990). The second value is 36 times larger than the first!

In addition to the number of accidents, the number of miles traveled (exposure data) is needed. The most commonly cited sources of truck travel data and their strengths and weaknesses are shown in Table 3-2. The exposure data also suffer from a lack of consistency. The TIUS and HPMS discrepancy was about 25% in 1982 (the last year both were available). Similarly, in 1985, the HPMS and the NTTIS discrepancy was about 50% (TRB 1990). In each case the HPMS was higher.

In addition to the variation found within the data for each of the two parameters (the number of accidents and the number of miles traveled), the data for each do not correspond well with each other because the two types of data are collected by independent mechanisms. Accident data are based on police reports, and exposure data are based on questionnaires or traffic counts. A partial listing of the factors thought to have an influence on truck accident rate is contained in Table 3-3 (Harwood and Russell 1990). The two types of data may have the additional problem that they currently cannot be disaggregated by the same factors, that is, a subset of those from Table 3-3. Because of this lack of correspondence, it usually is necessary to make

TABLE 3-1. Summary of existing truck accident data programs

Database	Agency	Time frame	Features and strengths	Limitations
FARS	NHTSA	Since 1975	Census of fatal crashes; has good quality control.	Does not include nonfatal crashes; limited detail on truck and operation.
TIFA	UMTRI	Since 1980	Census of fatal truck crashes; rich detail on truck, carrier, and driver; good quality control.	Does not include nonfatal crashes.
GES	NHTSA	Since 1988[a]	Based on a probability sample of fatal and nonfatal crashes.	Limited detail on truck and operation; based solely on police reports.
CARDfile	NHTSA	Since 1982	Large sample of fatal and nonfatal crashes.	Not based on a probability sample; limited detail on truck and operation; no road class; based solely on police reports.
Computerized Motor Carrier Accident Reports (50-T)	Office of Motor Carriers, FHWA	Since 1973	Rich details on trucks.	Lacks data on intrastate carriers; based solely on self-reporting.
SAFETYNET	Office of Motor Carriers, FHWA	Since 1990	Census of reported truck crashes (interstate carriers).	No truck configuration, driver, or crash type; based solely on police reports.
State accident data	Individual states		Census of reported fatal and nonfatal crashes.	Data elements and reporting thresholds not uniform among states; based solely on police reports.
NTSB reports	NTSB		Rich details on contributing factors.	Number of crashes investigated extremely small and not randomly selected.

Source: TRB 1990.

[a]Until 1988, the NHTSA National Accident Sampling System collected fatal and nonfatal truck crash data.

TABLE 3-2. Summary of existing truck travel data programs

Database	Agency	Data source	Strengths	Limitations
TIUS	Census Bureau	Questionniare survey of truck operators sampled from vehicle registration file.	Large sample size.	Does not capture current truck populations because of time lag between the sampling and survey years; reflects tractor use in 12 months, not the actual configuration; no road class, region, time of day, or month; based solely on self-reporting; available only once every five years.
NTACS	Census Bureau	Questionnaire surveys of TIUS subsamples.	Reflects trips made by the vehicle, not just the tractor; has road class and commodity flow.	Has greater time-lag problem than TIUS; based solely on self-reporting; available only once every five years.
NTTIS	UMTRI	Telephone interviews of truck operators sampled from vehicle registration file.	Reflects trips made by the vehicle, not just the tractors; has details on truck, road, environment, carrier, and driver.	Has time-lag problem similar to that in TIUS: based mostly on self-reporting; no state or region; available for 1985 only.
HPMS	FHWA	Traffic counts by states on an ongoing basis.	Reflects current truck populations; road class and region readily available; based on a probability sample of road sections.	Lacks details other than truck configuration, road, and region; lacks quality control.
ATWS	FHWA	Truck weighing by states.	Same as HPMS, but not based on a probability sample.	Not based on a probability sample; lacks quality control.

Source: TRB 1990.

TABLE 3-3. Partial listing of factors considered to affect truck accidents

Truck Type or Configuration	*Highway*
Number of trailers	Function
Number of axles on tractor/trailer(s)	Access control
Cab type	Number of lanes
Cargo area configuration	Lane width
Truck Size and Weight	Shoulder width
Width of trailer	Shoulder surface
Length, overall	Median width
Length, trailer(s)	Horizontal alignment
Empty/loaded	Vertical alignment
Weight, gross	Surface condition (wet/dry/etc.)
Weight, trailer	Pavement condition
Truck Operations	Pavement type
Cargo type	*Traffic*
Operator type	Volume (ADT)
Trip type	Volume (day/night)
Truck Driver	Percent trucks
Age	*Environment*
Experience with rig	Visibility
Hours of service	Weather
Driver condition	Light
Location	*Temporal*
State	Month/season of year
Urban/rural	Day of week
	Time of day

Source: Harwood and Russell 1990.

assumptions or to adjust the data to attempt to obtain corresponding values for the numerator and the denominator.

The available data are sure to improve in the middle to late 1990s. In 1989, the National Governor's Association recommended a set of uniform data elements that all states should collect. In 1990, the Transportation Research Board developed a plan to assemble a National Monitoring System (TRB 1990).

Given the large variations in reported values, the lack of consistency, and the lack of desired detail, what should one do to obtain reasonable accident rates? The analyst has three basic options. The first is to obtain one or more of the databases previously described and perform analyses to obtain both accident and travel data for the specific conditions desired, assuming that the databases permit the desired distinction to be made. Most authoritative accident rate studies have used this option; they have thoroughly examined the data to eliminate duplications, to estimate the extent of missing data by comparing individual entries in multiple databases, to correct coding errors, to determine data adjustment factors, and/or to disaggreate the reported values. The values obtained are by definition averages for the nation or the state(s) for which the data were collected, which then are used for the specific route(s) of interest. This is the most expensive and time-consuming option and involves appropriate statistical analyses (Hu et al. 1989; Chira-Chavala 1991). When examining data, the analyst must distinguish between: (1) truck and car data; (2) different types of trucks (e.g., tractor-trailer, tractor with two trailers, tank trucks, etc.); (3) the accident rate and the vehicle involvement rate; and (4) the estimated reporting efficiencies for different types of accidents (e.g., fatal vs. property damage only).

The second option is to access state database(s) for the specific route(s). States frequently have accident and travel data for the major state highways by milepost number. A necessary procedure is to perform a statistical test with expected values to determine if the calculated value should be replaced because the sample size is too small (Harwood and Russell 1990). The expected value could be used if data were missing for a highway segment. (The expected value(s) can be those from the third option.) This is the most defensible option to the general public because the data are for the exact route(s) of interest.

The third option is to use one of the existing analyses of the databases and apply the results to the specific route(s) of interest. This option is the least expensive and should result in values that are reasonably accurate considering all of the uncertainties in the other risk parameters (to be described later). Since at least the early 1980s, it has been known that the truck accident rate varies with road type and with population density. The most widely cited values (Harwood and Russell 1990) are shown in Table 3-4.

TABLE 3-4. Truck accident rates by state and combined

Highway class		Truck Accident Rate (accidents per million veh-mi)			
Area type	Roadway type	California	Illinois	Michigan	Weighted average[a]
Rural	Two-lane	1.73	3.13	2.22	2.19
Rural	Multilane undivided	5.44	2.13	9.50	4.49
Rural	Multilane divided	1.23	4.80	5.66	2.15
Rural	Freeway	0.53	0.46	1.18	0.64
Urban	Two-lane	4.23	11.10	10.93	8.66
Urban	Multilane undivided	13.02	17.05	10.37	13.92
Urban	Multilane divided	3.50	14.80	10.60	12.47
Urban	One-way street	6.60	26.36	8.08	9.70
Urban	Freeway	1.59	5.82	2.80	2.18

Source: Harwood and Russell 1990.
[a]Weighted by veh-mi of truck travel.

Harwood and Russell (1990) looked for databases that included data on highway geometrics, truck volumes, and truck accidents. Three states had computerized data files: California, Illinois, and Michigan. Table 3-4 lists the accident rate values for each of the three states and the weighted average obtained by summing the numerators and the denominators. The weighted average values in Table 3-4 are proposed by FHWA as national default values, that is, for use when better data are unavailable. The substantial variation in accident rate in Table 3-4 has been criticized because some state values given are nearly twice or are one-half of the average (Hobeika and Kim 1991); Harwood and Russell (1990) point out that the largest variations occur where the sample sizes were the smallest, and the weighted-average minimizes the effect of such values. They also refer to other studies with large state-to-state variations. Some differences do exist in state reporting requirements, but Harwood and Russell (1990) could come to no hard conclusions about the cause(s) of the large variations in Table 3-4.

A conclusion that can be reached at this point is that the values in Table 3-4 have an uncertainty of as much as $\pm$ 2. Numerous earlier studies resulted in accident rates that substantiate the values in Table 3-4 (Smith and Wilmot 1982; Abkowitz et al. 1984; Graf and Archuleta 1985; Jovanis et al. 1989). Generally, the highway type descriptors were somewhat variable, making some comparisons less than direct. It is common for all but one or two values to be very close (by less than a factor of two) to those in Table 3-4 and the remaining values(s) to be different by as much as a factor of five. These isolated, large variations are sometimes associated with contrary trends; for example, the rural freeway may have an accident rate greater than the urban accident rate, not the reverse. A set of values that are consistent with Table 3-4 and are aggregated in a way that is sometimes useful is presented in Table 3-5 (Jovanis et al. 1989).

All database analyses are not equally valid. For example, Moses and Savage (1991) derived an accident rate for hazardous material haulers of 7.8 accidents per million miles, compared to the FHWA value of 0.74 accidents per million miles (Grimm 1991) for the same database. The user of reported data analyses must develop a confidence level with the methodology used to

TABLE 3-5. Truck accident rates by highway type

Highway type	Accident rate
Controlled access	3.8×10^{-6}/mile
Noncontrolled access	28.4×10^{-6}/mile
Local streets	15.6×10^{-6}/mile

Compiled from: Jovanis et al. 1989.

analyze the data. Part of developing that confidence is to compare the results reported by several groups. Ideally the various researchers will describe any differences from previously reported values and explain the variations. In the case just cited, the difference appears to be the use by Moses and Savage (1991) of "total accidents" rather than only accidents above the regulatory reporting threshold. The increased potential for hazardous materials carriers to report minor accidents below the threshold contributed to the differences between the FHWA and Moses and Savage (1991) values.

Harwood et al. (1990) attempted to correlate accident rate with annual average daily traffic and the percentage of trucks; however, they obtained no consistent results. Other researchers have correlated the effect of other factors, such as day/night and wet/dry pavement (Saccomanno and Chan 1985). Differences in accident rate for these types of factors are small relative to the uncertainties in the other risk parameters (to be described later). Analysts' current means of estimating the effects of highway geometric conditions such as bridges, lane widths, and so on, on accident rate are not well understood quantitatively (TRB 1987a), but this is an active area of research (TRB 1987b). The current correlations in the literature are useful in some special cases (Neuman et al. 1991a,b).

In conclusion, the uncertainty in the truck accident rate values that can be obtained from a well-designed analysis of currently available databases is about a factor of two or three. Further, the ability to correlate truck accident rate values with factors that influence truck accidents is limited. If one is formulating a safety program that depends on data, for example, to distinguish between different pavement types and conditions, then the available data are inadequate. On the other hand, if one is planning the emergency response needs of a community or even a state, then the data are quite adequate (FEMA et al. undated). As will be discussed elsewhere in this book, the uncertainty in the truck accident rate is small compared to the uncertainty in some of the other parameters used in absolute risk calculations. Even relative risk calculations also may incorporate other parameters with high uncertainty. For many risk calculations, the data in Table 3-4 are reasonable; so expending resources to derive other values may not be cost-effective. The interested reader is referred to Harwood and Russell (1990), TRB (1990), and Abkowitz and List (1986) for detailed critiques of the available databases.

3.1.2 Train

Train accident rates are more straightforward than those for trucks because there is basically one source of accident data for a train and one source of exposure data. The Railroad Accident/Incident Reporting System (RAIRS) is a database of accident reports submitted by the railroads to the Federal Railroad Administration (FRA). The FRA has a dedicated staff that inputs the data from accident forms and performs a number of internal consistency checks. Accident reports are filed when (1) there is any impact with rail equipment and highway users at a rail-highway crossing, (2) the operation of the railroad results in death or injury, and (3) any accident results in damages exceeding the reportable threshold. From 1962 to 1974 the damage threshold was $750. Since then, the threshold has been changed biannually to reflect inflation. The value for 1991 and 1992 ($6300) was retained for 1993. A change for 1994 had not been established at press time.

The Interstate Commerce Commission (ICC) requires the railroads to submit a portion of the waybills from the Class I carriers, that is, the major railroads. Every tenth waybill showing origin, destination, commodity, number of cars, railroad junctions traversed, and so on, is submitted. In 1981, the sample size was increased to about 6% by requiring a higher sampling rate for multiple-carload shipments. The Association of American Railroads (AAR) took the responsibility in 1983 for collecting the waybills and preparing the data submitted to the ICC. The AAR has introduced numerous cross-checks and has worked with the railroads to improve the quality of the database (Abkowitz and List 1986).

The AAR also maintains a database (TRAIN II) on railcar movements. Of the railcars included, 100% of their movements and the commodity being carried are recorded; however, only about 90% (1991) of all railcars are included (BOE 1992a). The AAR Bureau of Explosives (BOE) uses TRAIN II data to compute nonaccident leaks per 1000 cars shipped and various commodity flow information (see Section 3.3.3).

The FRA Office of Safety (OS) publishes an annual accident/incident report, *Accident/Incident Bulletin*, which presents a wealth of information including: (1) train accidents by speed, reported cause, track class and type (e.g., derailment); (2) overall accident rate and accident rate by Class I railroad; and (3) train miles. One of many accident data tables from the report (FRA 1992) is shown in Table 3-6. FRA-OS also publishes an annual *Rail-Highway Crossing Accident/Incident and Inventory Bulletin.*

The train miles for the years 1986–91 are shown in Table 3-7, and accident totals for the same period are shown in Table 3-8. Using these two tables, various accident rates can be computed. The accident rate most often quoted, the rate excluding grade crossing accidents, is shown in Fig. 3-1. An accident rate for yards can be computed: 14.4×10^{-6}/mile for 1991. Unfortunately, there is not a one-to-one correspondence for the other values of Tables 3-7 and 3-8. The accident rate without the yard or crossing values resulted in an overall rate of 2.4×10^{-6}/mile for 1991. Because the FRA does not have train mileage by track class, an accident rate by track class cannot be easily computed.

TABLE 3-6. Train accidents and reportable damages by track class, 1991

Track class	Total accidents	Total damage[a]	Equipment damage[a]	Track damage[a]	Collisions	Derailments	Other accidents	Track caused	Equipment failure	Human factors	Other causes
?	235	6,564,459	4,944,117	1,620,342	28	166	41	100	11	83	41
1	1,291	39,196,284	28,870,153	10,326,131	130	936	225	517	94	510	170
2	443	25,158,594	18,059,419	7,099,175	32	324	87	178	58	122	85
3	342	46,810,944	36,073,682	10,737,262	28	234	80	84	98	71	89
4	384	89,044,945	69,084,737	19,960,208	29	213	142	67	102	81	134
5	90	10,267,699	7,813,115	2,454,584	5	63	22	14	34	18	24
6	29	5,894,070	4,706,214	1,187,856	9	0	20	8	18	2	1
TOTAL	2,814	222,936,995	169,551,437	53,385,558	261	1,936	617	968	415	887	544

Source: FRA 1992.
[a]Values are in dollars.

TABLE 3-7. Train miles, 1986 through 1991[a]

Year	Motor	Yard switching	Locomotive	Total
1986	0.340	1.082	4.248	5.671
1987	0.341	1.020	4.453	5.813
1988	0.349	1.053	4.691	6.093
1989	0.325	1.043	4.837	6.206
1990	0.332	0.982	4.774	6.088
1991	0.281	0.885	4.602	5.768

Source: FRA 1992.
[a]Values are in one hundred million miles.

TABLE 3-8. Number of train accidents, 1986 through 1991

Year	Mainline	Yard	Miscellaneous[a]	Crossings[b]	Total
1986	1292	1221	248	141	2761
1987	1292	1091	264	135	2647
1988	1424	1360	267	197	3051
1989	1404	1403	273	182	3080
1990	1298	1431	316	166	3045
1991	1267	1277	270	156	2814

Compiled from: FRA 1987-92.
[a]Industry, siding, and unknown. [b]Included in mainline, yard, or miscellaneous accidents.

Nayak et al. (1983) computed collision, derailment, and overall accident rates by track class by using RAIRS accident data by track class for 1975, 1976, and 1977 and using an estimate of train miles by track class using the 1976 ICC waybill data. The overall accident rate shown in Fig. 3-1 has declined significantly since 1976; so, to be of use, the track class data are normalized as shown in Table 3-9. Lacking better data, the values from Table 3-9 can be used as multipliers to the values from Fig. 3-1 (or Tables 3-7 and 3-8) to compute an estimated accident rate by track class. Track class 1 is the poorest, and 6 is the best. Classes 5 and 6 were combined by Nayak et al. (1983) owing to the small number of accidents.

In addition to the publicly available data and data analyses previously described, some industry organizations and some individual chemical companies possess proprietary databases. The AAR, the Railway Progress Institute (RPI), and the Chemical Manufacturers Association (CMA) are jointly developing a hazardous materials transportation risk analysis model. The accident cause portion of the model will include the effects of accident type (collision or noncollision), train speed, track class, railroad type (e.g., Class I), train length, car placement within the train, tank car type, and car protective features. The accident cause model is the subject of major research

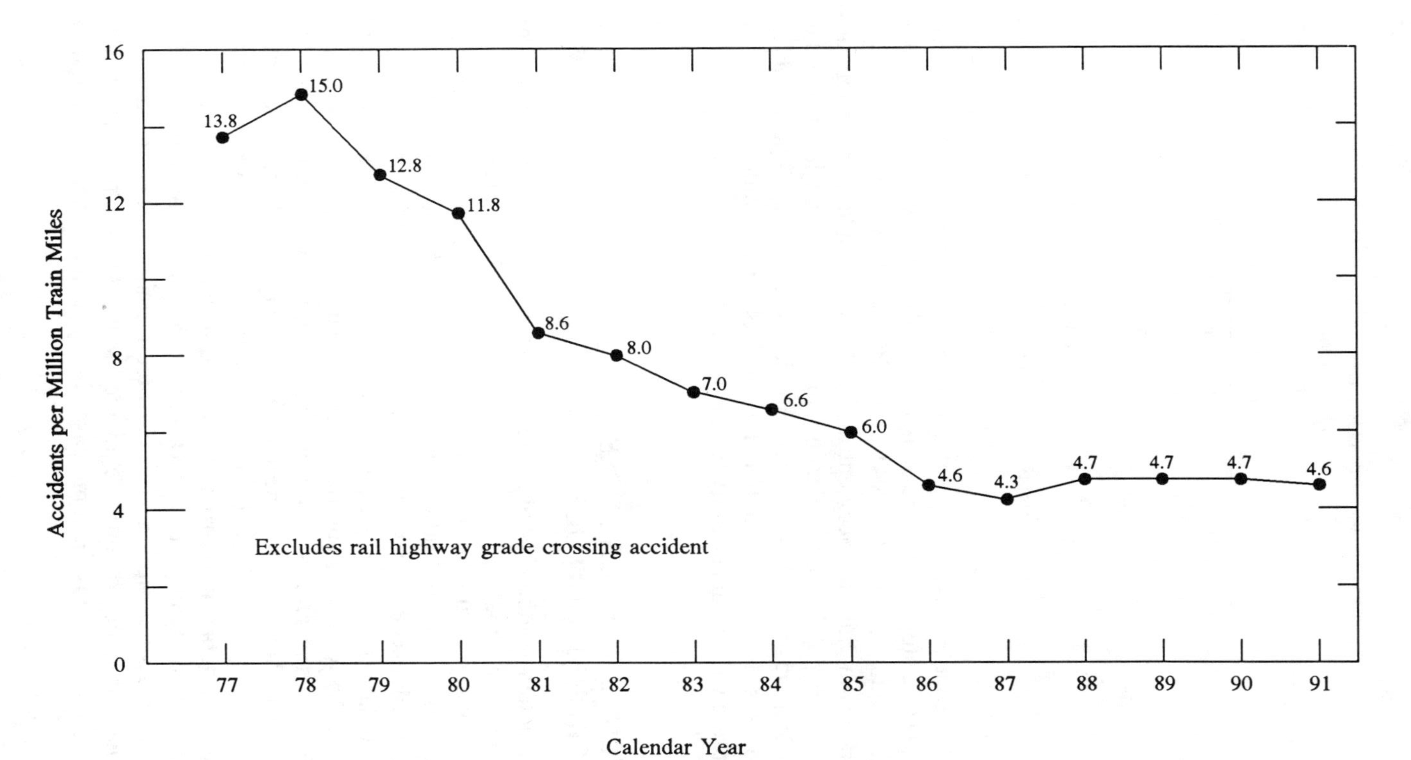

FIGURE 3-1. Railroad accident rate trends, 1977 through 1991. Adapted from: FRA 1987-92 and RSPA 1988.

TABLE 3-9. Accident rate adjustment factor for track classes

Track class	Adjustment factor
1	21.0
2	6.7
3	2.2
4	0.25
5/6	0.37
Average	1.0

Adapted from: Nayak et al. 1983.

efforts; thus, one can expect a great deal of new information (Bendixen and Barkan 1992). Unfortunately, the analytical model is likely to be considered proprietary to members of the AAR, the RPI, and the CMA.

In conclusion, the train accident rate and many accident characteristics are reasonably well documented. A notable exception is the data needed to directly compute an accident rate as a function of track class. The use of Fig. 3-1, or Tables 3-7 and 3-8, together with Table 3-9, should produce an accident rate with an uncertainty factor of about two.

3.2 ACCIDENT FORCE TYPES AND FORCE MAGNITUDES

Two parameters needed for a quantitative transportation risk analysis using the predictive methodology (Section 2.4) are described in this section: (1) the conditional probability that, given an accident, a particular force type (e.g., impact) will occur; and (2) the conditional probability that the force magnitude will exceed a certain value. The two parameters are presented together in this section because the source of data for both parameters is the same, the bases for any assumptions for each parameter type frequently are the same, and the results of the analysis frequently are intertwined for presentation and ease of use. The accident force type probability is derived directly from accident data; hence, it is relatively easy to compare the original 1970s data with current data. The accident force magnitude is determined by combining data (e.g., accident speeds) with analytical models (e.g., spring-mass description of truck tractor, trailer, and cargo striking a hard surface) that are not always explained well enough to permit independent computation. Thus, in some cases it is possible to verify initial data and/or assumptions but impossible to recompute final results on the basis of new data or assumptions.

The original analyses were made for a truck trailer or railcar with a load of small radioactive material containers (Clarke et al. 1976), but the analyses should be applicable to typical hazardous material containers, e.g., 55-gal drums. A subsequent analysis was based on large, radioactive materials containers but is applicable to the even larger bulk containers typical of hazardous materials (Dennis et al. 1978). Unless otherwise specified, the bulk transport data and analyses are from Dennis et al. (1978), and the small container data and analyses are from Clarke et al. (1976). In both cases, the researchers were from the Sandia National Laboratories (SNL) and were funded by the DOT and the Department of Energy (DOE). In this book Clarke et al. (1976) and Dennis et al. (1978) sometimes are referred to generically as SNL because much of the basic data are the same, and some of the analytical models are the same.

Five accident environments were defined and investigated: fire, impact, crush, puncture, and immersion. The fire temperature from hydrocarbon fuels varies from 1400°F to 2400°F; an 1850°F fire temperature was considered representative. Impact is defined as striking or being struck by an object that has no sharp projections. Crush is characterized by structural loads, either highly localized or over a large area, that cannot be described as impact or puncture. Puncture is defined as striking or being struck by an object that has the potential to penetrate the container. Immersion is not considered a threat to most containers; therefore, it is not presented in this book.

The data and analysis summaries that follow are intended to present data that can be used by the risk analyst and to convey an understanding of the assumptions they are based on. The interested reader is referred to the chapter-end references for more detail.

3.2.1 Bulk Transport By Truck

Section 3.2.1 is based on Dennis et al. (1978) unless specified otherwise.

3.2.1.1 Fire Force

In 1986, researchers at the Lawrence Livermore National Laboratory (LLNL) concurred with the complete SNL fire model after an independent review of the data then currently available (Fischer et al. 1987). Fires occur in approximately 1.6% of all truck accidents. The second column of Table 3-10 shows the conditional fire frequency, given the accident type in column one, and the third column shows the corresponding percentage fraction of all fires.

TABLE 3-10. Truck accident fire distribution

Accident type	Conditional fire frequency	Percent of total fires
Collision	0.004	24.9
Overturn	0.012	2.1
Run off road	0.011	4.9
Fire only	1.000	19.3
Other noncollision	0.130	48.8
		100.0

Source: Clarke et al. 1976.

The fire duration distribution given in Fig. 3-2 is based on a statistical analysis process that incorporates a number of assumed probability distributions. Typical assumed distributions are given here for illustration:

1. A truck/truck collision has a 2% chance that one or more of the trucks is a tanker carrying flammable goods.
2. A truck/tanker collision has available fuel that can vary in quantity from zero to 10,000 gal; the expected value is 5,000 gal. The fuel cumulative distribution function (cdf) is given by the expression:

$$1 - e^{-(f/5718)^{2.73}} \tag{3-2}$$

   where f is fuel in gallons.
3. No or minimal fire-fighting capability is available for 60% of accidents, a local fire department is summoned for 30% of accidents, and hand extinguishers are the only means used 10% of the time.
4. Local fire departments are effective against all fires. Unless a fire burns out before they arrive, the fire department limits the fire duration to between 15 and 45 min, with a uniform distribution.

If the risk analysis is for a tank truck containing flammable material, then assumptions 1 and 2 are neither appropriate nor conservative.

Sufficient data are not available to establish how the bulk container or truck tank is oriented with respect to the fire resulting from an accident. A conservative analysis approach is taken to analyze the container response to an enveloping fire with a temperature of 1850°F, to determine the fire duration required for container failure. Figure 3-2 then can be used to determine the probability of a fire of sufficiently long duration. Fischer et al. (1987) assumed that the bulk container is sideways to a fire located from zero to 31.5 ft away. The fire distance distribution was assumed to be equally

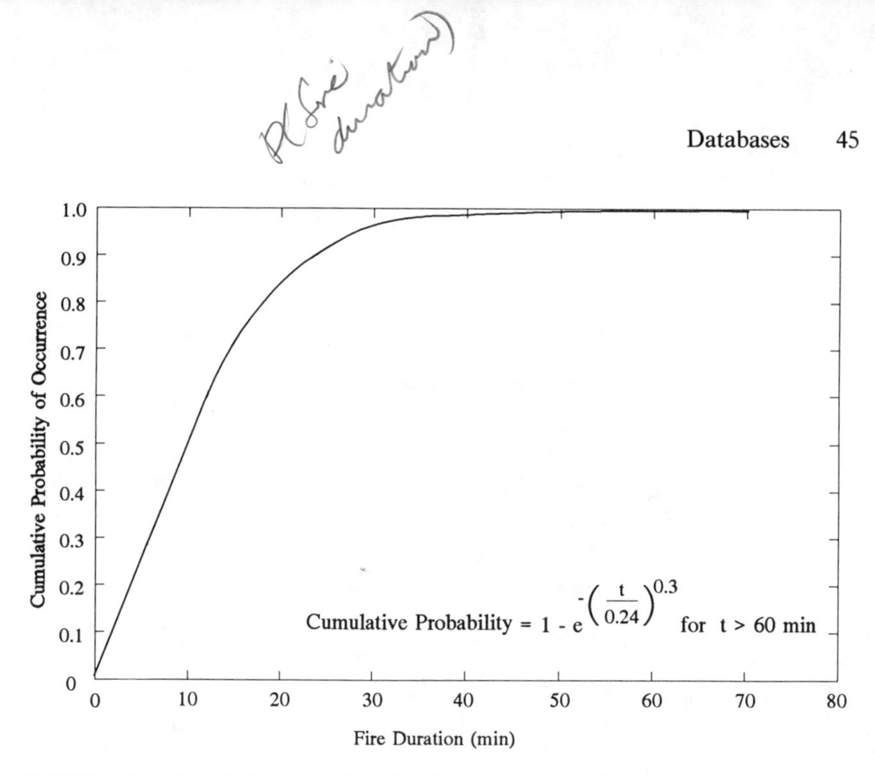

FIGURE 3-2. Cumulative probability distribution of fire-accident duration for truck transport of large containers. Source: Dennis et al. 1978.

likely in this interval. Using a fire distance of other than zero feet implies multiple fire analyses, and most risk analysts use the enveloping fire model for simplicity and conservatism.

3.2.1.2 Impact Force

Accidents can be considered to be collision or noncollision. For bulk containers, Dennis et al. (1978) examined the 1969 to 1972 accident distribution data to determine a fraction for collision with relatively hard objects as the only ones that would threaten the container. The result was 80.2% "collisions" and 19.8% "noncollisions." Fischer et al. (1987) used data from 1973 through 1983 and obtained 74.12% collisions, 24.91% other mechanical loading, and 0.97% fire only (Fig. 3-3). If we delete accident numbers 1, 2, and 17, the collision with hard objects percentage is 69.4. If we delete accident numbers 19, 22, 25, 29, and 31 as probably not challenging the container, then the percentage of noncollision accidents into potentially hard objects is reduced to 10.9%. The total hard impacts is then 80.3%, which is in close agreement with the Dennis et al. (1978) value. Harwood and Russell (1990) reported collision type data for tractor-single trailer combinations for 1984 and 1985. Their values are aggregated and presented in Table 3-11 for comparison with the Fischer et al. (1987) values. Although there are some small differences, they are not sufficient to invalidate the Fischer et al. (1987) results and hence the SNL results.

Truck accident					Probability percent **	Accident index
Collision 0.7412	Nonfixed object 0.8805	"Soft objects" cones, animals, pedestrians 0.0521			3.4002	1
		Motorcycle 0.0124			0.8093	2
		Automobile 0.6612			43.1517	3
		Truck, bus 0.2041			13.3201	4
		Train 0.0118			0.7701	5*
		Other 0.0584			3.8113	6
	On road fixed obj. 0.1195	Bridge railing 0.0577	Water 0.20339		0.1039	7*
			Railbed / roadbed 0.77965		0.3986	8*
			Clay, silt 0.015486		0.0079	9*
			Hard soil, soft rock 0.001262		0.0006	10*
			Hard rock 0.000199		0.0001	11*
		Column, abutment 0.0042	Column 0.9688	Small 0.8289	0.0299	12*
				Large 0.1711	0.0062	13*
			Abutment 0.0382		0.0011	14*
		Concr. obj, bottom str. 0.0096			0.0850	15
		Wall barrier, wall, post 0.4525			4.0079	16
		Signs, cushions 0.0577			0.5111	17
		Curb, culvert 0.4183			3.7050	18
Noncollision 0.2588						

* Potentially significant accident scenarios
** Conditional probability which assumes an accident occurs

FIGURE 3-3. Truck accident scenarios and their percent probabilities. Source: Fischer et al. 1987.

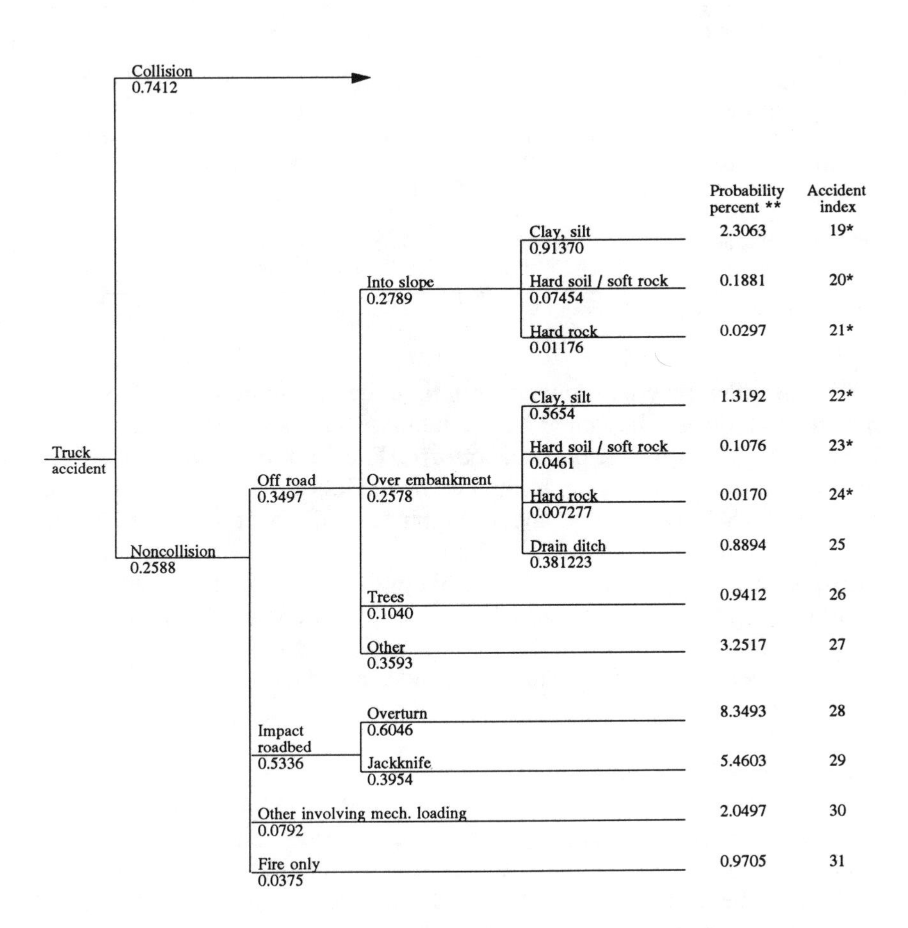

FIGURE 3-3. (*Continued*)

TABLE 3-11. Comparison of truck accident types

Category	Hardwood and Russell (1990)	Category	Fischer et al. (1987)
Collision with parked vehicles and moving car	47.9	Collision with car	43.2
Collision with moving truck	11.2	Collision with truck/bus	13.3
Collision with train	0.5	Collision with train	0.8
Other collision	17.2	Other collision	16.8
Total collision	76.8	Total collision	74.1
Fire	0.6	Fire only	1.0
Ran off road	6.2	Off road	9.0
Other noncollision	16.2	Other noncollision	15.9
Total noncollision	23.0	Total noncollision	25.9

Compiled from: Harwood and Russell 1990 and Fischer et al. 1987.

The accident severity (impact magnitude) determination by SNL is summarized as follows. The same 1969-72 database was used to determine if the impact was head-on, side-on, and so forth. The data for cumulative speed distribution were linearly approximated by the following: 0, 72.16 mph at a cdf of 97%, 80 mph at 99%, and 100 mph at 100%. A comparison of this distribution with that used by LLNL indicates that the SNL cumulative speed distribution is conservative (by 50% at 20 mph, 30% at 40 mph, and 20% at 50 mph) compared to that used by LLNL. The LLNL distribution is based on 1958 to 1967 California and 1979 to 1981 North Carolina data. Vehicle collisions were assumed to be inelastic, and statistical analysis of this model produced Table 3-12.

3.2.1.3 Crush Force

Crush is defined as the static force imposed by the truck lying on top of the container. The conditional probability of crush forces arising, given a collision or overturn accident, is conservatively estimated as 0.05. This value represents the probability that the truck trailer bed and the container are coincident after an accident. The conditional probability of collision or overturn accidents is 0.86; thus, the conditional probability of crush forces occurring, given any truck accident, is 0.043. On the basis of the trailer and ancillary equipment weighing no more than 15,000 pounds, the two loading configurations examined resulted in a maximum crush loading of slightly less than 2,000 pounds per lineal foot of loaded area. This model was developed for a large container on a flatbed trailer; for a tank truck, the conditional probability of a crush force due to the weight of the undercarriage, given an overturn accident, is 1.0. Thus, a better approximation for the

TABLE 3-12. Large container velocity change due to impact in a highway tansportation collision accident

Velocity change due to impact (mph)	Cumulative fraction of sample with a velocity change less than or equal to indicated velocity change for each over-the-road transport vehicle weight (in tons)						
	10	15	20	25	30	40	50
5	0.5038	0.6220	0.7146	0.7813	0.8248	0.8711	0.8962
10	0.7541	0.8489	0.8881	0.9104	0.9256	0.9454	0.9579
15	0.8759	0.9173	0.9386	0.9520	0.9610	0.9721	0.9782
20	0.9244	0.9508	0.9642	0.9723	0.9775	0.9834	0.9872
25	0.9528	0.9997	0.9782	0.9828	0.9859	0.9899	0.9923
30	0.9701	0.9810	0.9859	0.9890	0.9911	0.9936	0.9949
35	0.9809	0.9876	0.9910	0.9931	0.9945	0.9959	0.9966
40	0.9878	0.9921	0.9945	0.9950	0.9965	0.9973	0.9978
45	0.9922	0.9951	0.9966	0.9973	0.9977	0.9983	0.9987
50	0.9951	0.9970	0.9978	0.9983	0.9986	0.9990	0.9994
55	0.9970	0.9982	0.9987	0.9990	0.9996	0.9996	0.9998
60	0.9981	0.9989	0.9992	0.9995	0.9998	0.9998	0.9999
65	0.9989	0.9994	0.9996	0.9997	0.9999	0.9999	
70	0.9993	0.9996	0.9998	0.9999			
75	0.9996	0.9998	0.9999				
80	0.9997	0.9999					
85	0.9998						
90	0.9999						

Source: Dennis et al. 1978.

probability of a crush force occurring for a tank truck is the probability that the tank truck is overturned 180 degrees, given an accident. For an overturned tank truck and for most containers on a flatbed truck, crush is seldom a threat to the container or the tank.

3.2.1.4 Puncture Force

The SNL researchers considered the puncture environment the most difficult to estimate for both the truck and the train modes for both small and large containers. The analysis is considered as having "order-of-magnitude" applicability. The highway puncture threat is based on the rail threat described in Section 3.2.3.4, in which the railcar coupler is the primary puncture mechanism. For lack of a better assumption, the probe assumed for analysis is similar to a railcar coupler. It also is assumed that puncture would not occur for collision accidents with automobiles, light stationary objects, or terrain features. The vehicle speed distribution is the same as that discussed in Section 3.2.1.2. The analysis results are shown in Table 3-13; note that the conditional probability of a crush force has been incorporated into the table.

TABLE 3-13. Probability that a large container will be punctured during a truck accident

Package wall thickness (in.)	Probability per reportable accident
0.4375	1.77×10^{-3}
0.50	1.74×10^{-3}
0.75	1.63×10^{-3}
1.00	1.31×10^{-3}
1.25	7.50×10^{-4}
1.50	2.30×10^{-4}
1.75	3.18×10^{-5}
2.0	1.85×10^{-5}

Source: Adapted from Dennis et al. 1978.

3.2.2 Truck Transport of Small Containers

Section 3.2.2 is based on Clarke et al. (1976) unless specified otherwise.

3.2.2.1 Fire Force

The small container fire model is very similar to the large container fire model discussed in Section 3.2.1.1. The only difference is in the added assumptions in the statistical analysis model that the cargo of interest is nonflammable, but there is a 50% probability that it is shipped in a trailer with flammable cargo. The solid line in Fig. 3-4 shows the results. The error bars show the effects of changing by ± 50% the assumed fraction of trucks transporting flammable cargo, the amount of flammable material involved, and the fire-fighting effectiveness. The smaller error bars show the maximum and minimum values from changing the values individually; the larger error bars show the effect of changing the parameters to minimize or maximize the combined effect. The dashed line shows the effect of a truck carrying only nonflammable cargo. The dashed line is nearly the same as the line in Fig. 3-2.

The probability that a fire occurs, given an accident, is 1.6%, as in Section 3.2.1.1. Not every fire will result in exposure of the cargo to the fire. On the basis of limited data, Clarke et al. (1976) assumed that 40% of all fires following a truck accident would reach the cargo. Thus, given a truck accident, the probability that the cargo is exposed to the fire is estimated as 0.0064. This is the simplest approach. The recommended approach is to model the truck trailer as being enveloped by a fire of 1850°F temperature. The thin aluminum skin will soon melt in such an analysis, exposing the cargo to the fire. In a full truckload of containers, it is possible to compute the time to failure, and the associated probabilities, for each layer of cargo.

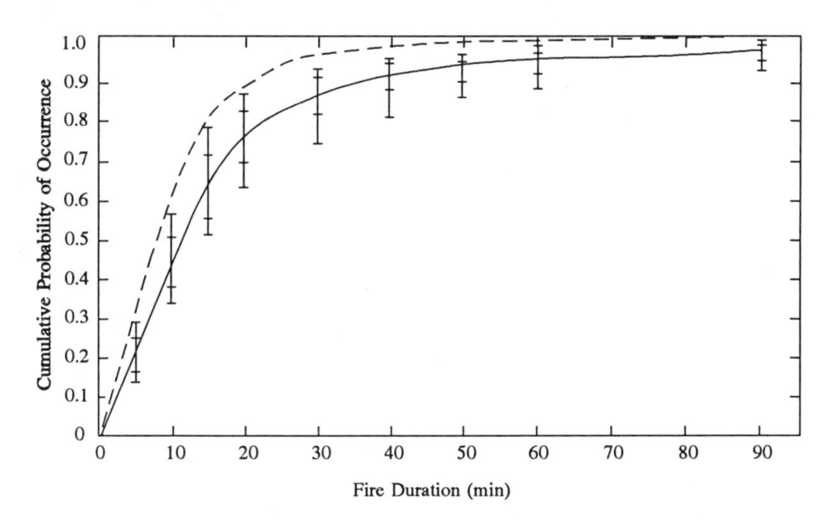

FIGURE 3-4. Cumulative probability distribution of fire-accident duration for truck transport of small containers. Source: Clarke et al. 1976.

One alternative to the recommended approach is to use the fire location distribution described in Section 3.2.1.1. A third, very conservative approach is to analyze the response of an individual container to an enveloping fire.

3.2.2.2 Impact Force

Clarke et al. (1976) assumed that the threat to the container is represented by the collision of the container with the trailer wall. The premise was that increased target rigidity due to tractor crushup does not occur until after container/trailer wall impact. The relative velocity between the container and the trailer wall was determined by a spring-mass model of the tractor-trailer combination as it strikes bridge abutments, and so on. The relative velocity result is shown in Fig. 3-5, and the energy absorption by the container is shown in Fig. 3-6. The impacted surface is the trailer wall. Containers frequently are tested by dropping them onto a hard surface; therefore, Clarke et al. (1976) converted Fig. 3-6 to an equivalent drop height as shown in Fig. 3-7. In this case the conditional probability of the impact force, given an accident, is the conditional probability of an overturn or collision, given an accident, or 0.89.

Given a truck accident, there is a probability of 0.0043 that the truck will be struck by a train. If the trailer is full, this is the probability of at least one container being struck. For one container of interest on the truck, the probability is 7.7×10^{-4} that it will be struck, given a truck accident. The

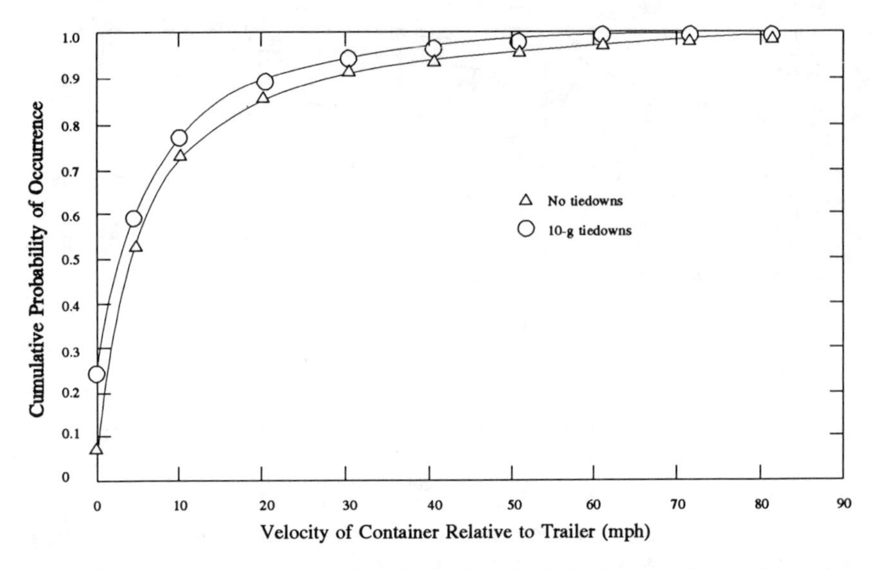

FIGURE 3-5. Cumulative probability distribution of relative impact velocity of container against trailer wall, given an impact accident. Source: Clarke et al. 1976.

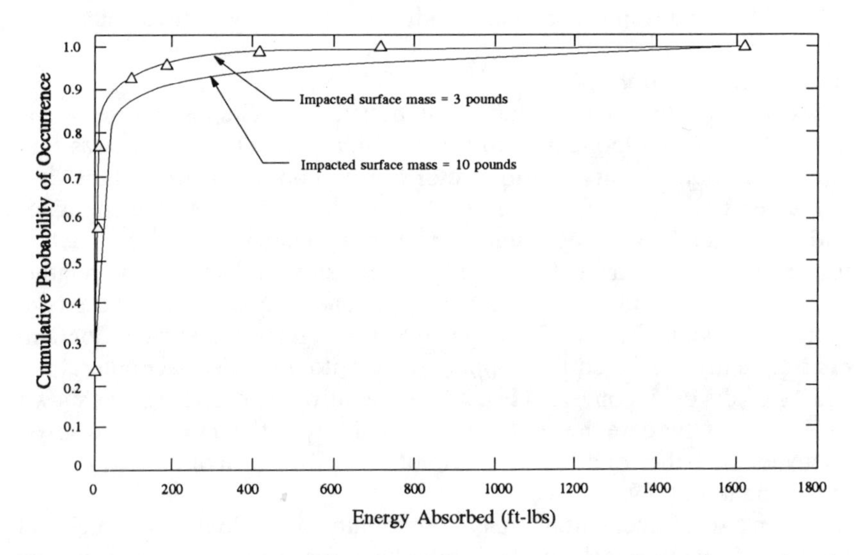

FIGURE 3-6. Cumulative probability distribution of energy absorbed by the container for a truck impact accident. Source: Clarke et al. 1976.

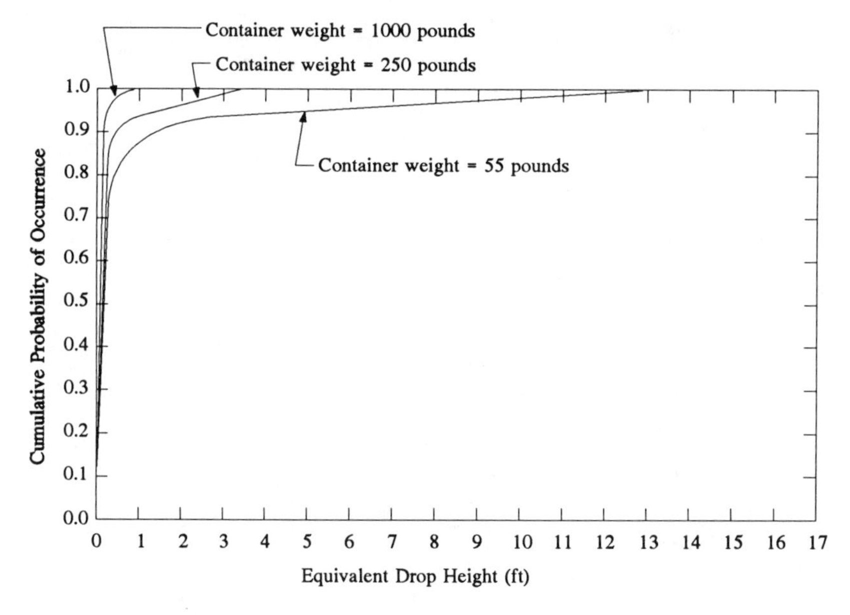

FIGURE 3-7. Cumulative probability distribution of occurrence of accidents with impacts equivalent to given drop height or less (assumes impacted surface mass is 3 pounds). Source: Clarke et al. 1976.

errors introduced by omitting the contribution of truck-train collisions are small when compared to the much more likely truck accident types.

3.2.2.3 Crush Force

Two types of crush forces were examined: (1) static forces due to the container being underneath the truck and (2) dynamic forces resulting from the transfer of momentum between containers in an accident. The static force model is similar to the large container model (Section 3.2.1.3) and was found to be insignificant compared to dynamic crush.

Dynamic crush was based on a truckload of rigid circular containers in contact with each other. Using the spring-mass model described in Section 3.2.2.2 to determine deceleration values for different types of collisions and considering that the containers in the front row experience more force than those in the back row leads to the cumulative probability distribution given in the second column of Table 3-14. Note that (1) the conditional probability of crush forces occurring has been incorporated, and (2) this distribution is

TABLE 3-14. Probability of a container experiencing a total crush force more severe than listed, given a truck accident

Crush force (lb)	Full load of rigid containers	Combined full and mixed loads
2,500	5.48×10^{-1}	2.46×10^{-1}
5,000	3.86×10^{-1}	1.17×10^{-1}
10,000	2.63×10^{-1}	5.20×10^{-2}
15,000	1.99×10^{-1}	3.13×10^{-2}
20,000	1.51×10^{-1}	1.88×10^{-2}
25,000	1.10×10^{-1}	1.16×10^{-2}
32,500	9.20×10^{-2}	4.48×10^{-3}
40,000	6.60×10^{-2}	3.56×10^{-3}
50,000	4.64×10^{-2}	2.68×10^{-3}
75,000	2.84×10^{-2}	1.80×10^{-3}
100,000	1.59×10^{-2}	8.80×10^{-4}
150,000	3.20×10^{-3}	3.56×10^{-5}

Source: Clarke et al. 1976.

the complement to those presented earlier in this chapter, that is, force more severe than indicated.

Given that some containers have failed as determined by Table 3-14, the risk depends on how many have failed. The model used to generate Table 3-14 is based on the force on the first row of drums being proportional to:

$$\frac{2N}{N_2} - 1 \tag{3-3}$$

with the second row force proportional to:

$$\frac{2N}{N_2} - 3 \tag{3-4}$$

and so on. The total number of containers is N, and N_2 is the average number of containers across the truck trailer width.

If some of the cargo is less rigid than the containers of interest, then reduced forces will occur on the rigid containers. The forces developed are highly dependent on the distribution of rigid and nonrigid containers within the truck trailer, but in such a situation the distribution is probably random from shipment to shipment. If we assume that 5% of the time a full truckload of rigid containers results and that 95% of the time one-fifth of the cargo is in rigid containers, then the third column of Table 3-14 results.

3.2.2.4 Puncture Force

The puncture model is based on data from 1971 and 1973 for accidents involving hazardous cargo, excluding tank trucks, which showed that in 31.1% of the accidents producing damage, the damage was due to puncture. Further, given a puncture in a container about the size of a 55-gal drum, about 6.8% of all the containers are punctured. The impact velocity model in Section 3.2.2.2 showed that about 72% of the accidents will involve relative velocity between the cargo and the trailer, and it is conservatively assumed that some cargo damage occurs, given a relative velocity difference. The puncture model is then:

$$0.8935\,\frac{\text{collisions or overturns}}{\text{truck accident}} \times 0.72\,\frac{\text{cargo damage accidents}}{\text{collision or overturn}} \times$$
$$0.311\,\frac{\text{puncture accidents}}{\text{cargo damage accident}} = 0.200 \text{ puncture accidents/truck accident} \tag{3-5}$$

If the number of containers of interest is less than 6.8%, then the conditional probability becomes 0.068 times 0.200 $\doteq 1.36 \times 10^{-2}$ punctures/container/truck accident. These results are for relatively soft containers that are easily punctured.

The hard container puncture model considered: (1) multiple puncture sources within and without the trailer (e.g., other cargo, floor, front wall); (2) assumed probability distributions for impact on these sources; (3) three different assumed distributions for the radius of the probe that might result from these sources; and (4) estimates of whether the probe would bend or penetrate. This information was combined with a container velocity model more sophisticated than the one used in Section 3.2.2.2, which produced an upper bound. The container velocity at impact is strongly dependent on the cargo spacing, that is, the distance of movement before impact occurs. An assumed distribution was used that is probably conservative: 20% move less than 3 in., 25% move more than the 2 ft distance required to reach maximum velocity, and the remainder fall in between these two extremes. Figure 3-8 presents the results when the various factors are combined using a statistical analysis process. The conditional probability of 0.20 puncture accidents/truck accident (or 1.36×10^{-2} on a container basis) needs to be used with Fig. 3-8.

A puncture probe must have a velocity to radius (V/R) value of about

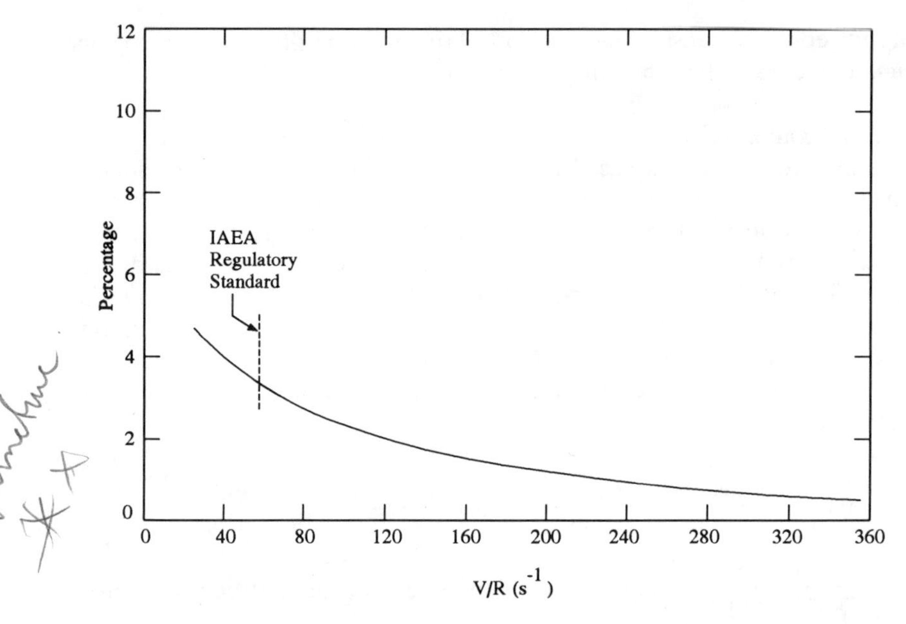

FIGURE 3-8. Percentage of puncture accidents for soft containers that would have resulted in puncture if containers had been hardened to the velocity to radius *(V/R)* value indicated. Source: Clarke et al. 1976.

50 s^{-1} in order to puncture a 17H steel drum. Using Fig. 3-8, the puncture failure probability is 0.037 punctures per puncture accident times 0.20 puncture accidents per truck accident or 7.4×10^{-4} punctures, given a truck accident. Comparison of this value to 0.20 punctures per accident for soft containers indicates that a steel drum is about 27 times less likely to be punctured than the average soft container.

3.2.3 Bulk Transport by Train

The most common train accident is a derailment, in which a car separates from the preceding car and leaves the track. A number of the cars behind the lead car follow that lead car. At the initial separation, the brake system for all cars is activated, and both segments of the train are stopped; thus, the derailed cars leave the track at successively lower speeds. The lead car is decelerated by impact with the terrain, and the following cars may impact a previously derailed car and potentially override it. The result is usually a jumble of cars lying parallel, across, and at odd angles to each other along a section of track. If a fire occurs, about 10 cars in the vicinity usually are affected. Train analyses are more complicated than truck analyses because the probability that the derailed segment includes the hazardous material of interest must be considered.

In a collision, the cars nearest the impact experience the highest velocity change, but the differential velocity change is rapidly reduced away from the point of impact. Typically, collision velocity changes for affected cars are greater than derailment velocity changes for affected cars. Section 3.2.3 is based on Dennis et al. (1978) unless specified otherwise.

3.2.3.1 Fire Force

The SNL conditional probability of fire is based on 1972 data showing that the average number of cars involved in a fire following a collision or derailment is 10, and the average number of cars involved in a fire following "other" accidents is one. Approximately 1% of the collision and derailment accidents studied resulted in fire, compared to 90% of other accidents. Combining these values with the number of accidents of each type in 1972 and dividing by the 1972 train miles and the average number of cars per train (66), SNL obtained a value of 2.8×10^{-8} car fires/car mile. This value is not useful because it is based on the relatively high 1972 train accident rate of 1×10^{-5}/train mile. (See Fig. 3-1.) Alternatively, the analysis described previously produces 0.184 car fires/train accident. Like the truck fire model, the SNL train fire model was reviewed in 1986 (Fischer et al. 1987). The conclusion was that no better estimate of fire probability or duration could be developed from the data then available.

The 0.184 car fires/train accident value does not take into consideration where the car(s) of interest are with respect to the fire. In 1972, about half of the cars in which fire occurred were involved in derailments or collisions affecting an average of 10 cars. The other half of the cars were involved in other types of accidents in which only one car was involved. As a first approximation, the author suggests that the fire rate stated here be adjusted for one car of interest by a factor of $(1 + 10)/2N$, where N is the number of cars in the train. The first factor is for single-car fires and the second for multiple-car fires.

An important consideration in the analysis of tank cars involved in fire is whether the tank car is oriented so that liquid or vapor is released from valves. One-half of the fires in 1972 were from collisions and derailments. Data presented in Section 3.2.3.2 show that at least 62% of all collision and derailments involve car rollover. The other half of fire accidents are one-car accidents, and the cars are probably upright on the track. As a first approximation, the author recommends that $(1)(0.5) + (1 - .62)(0.5) = 0.69$ be used as the fraction of cars upright in a fire, and 0.31 be used as the fraction of cars overturned in a fire.

The fire duration or severity model is similar to that used for truck

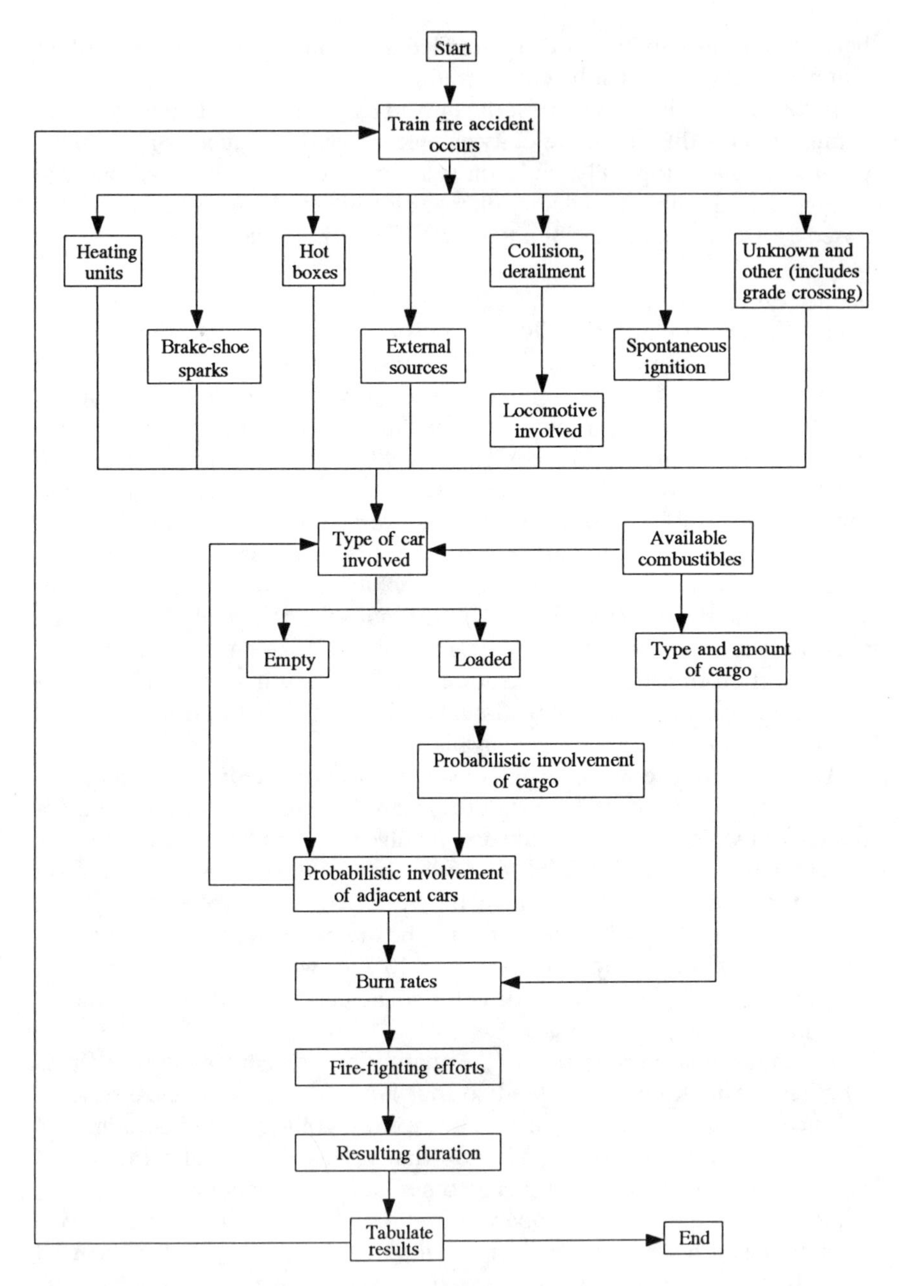

FIGURE 3-9. Simplified flow chart of the train-fire-duration program. Source: Dennis et al. 1978.

accidents: a statistical analysis of assumed probability distributions for sources and amounts of combustible materials and fire-fighting efforts (Fig. 3-9). Typical assumed distributions are given here for illustration:

- Loaded tank cars are involved in 14% of the fires.
- When one car becomes involved in a fire, the probability of adjacent cars (if flammable) catching fire is 20%.
- Loaded tank cars carry 100% flammables, with capacities ranging from 10,000 to 40,000 gal and a cdf given by:

$$1 - e^{-(C-10,000/17,842)^{2.11}} \tag{3-6}$$

where C is capacity in gallons.

The results of the statistical analysis are presented in Fig. 3-10. The effect of requiring shelf couplers, head shields, and thermal shields to be retrofitted and put on many new cars carrying hazardous materials will make this analysis more conservative.

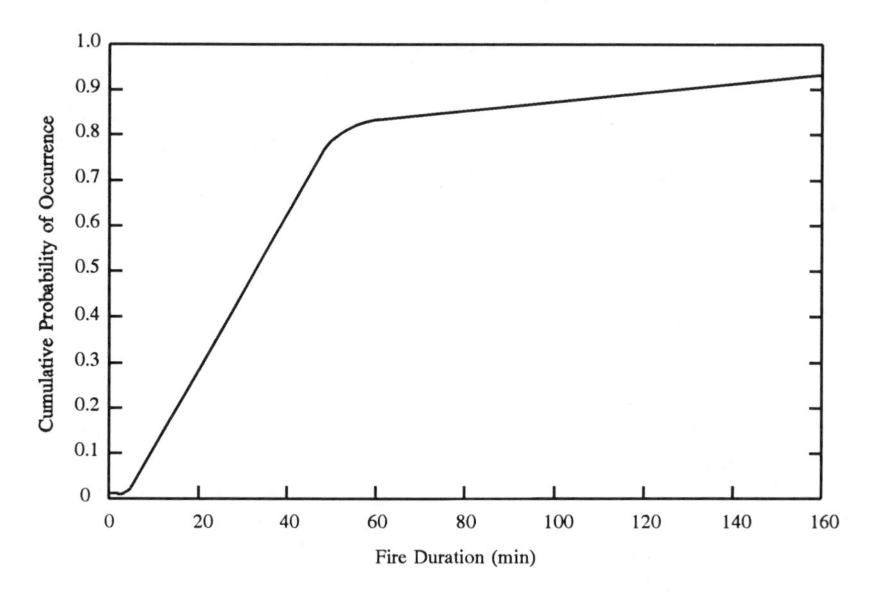

FIGURE 3-10. Cumulative probability distribution for duration of train-fire accidents. Source: Dennis et al. 1978.

3.2.3.2 Impact Force

The conditional probability of impact is the probability of collision or derailment, 0.90. To determine the impact force magnitude, a statistical analysis was performed by SNL for both derailments and collisions. A derailment speed distribution and two collision speed distributions (routine and broken train) were based on 1973 FRA data. The LLNL analysis was based on a single collision speed distribution and three derailment speed distributions (routine, over bridges, and over embankments). The LLNL speed distributions are based on 1979 through 1982 data. The routine LLNL derailment speed distribution and the SNL derailment speed distributions are within 5% of each other. The LLNL collision speed distribution and the SNL routine collision distribution are within 2% of each other at 50 mph. At lower speeds, the LLNL distribution is consistently lower by increasing amounts. The LLNL distribution is more conservative, having one-third fewer accidents in the zero to 10 mph range and more in the 30 to 50 mph range.

Given a derailment speed, SNL modeled the effective velocity change as a linear distribution from 100% to zero to account for hardness of surface struck, angle of impact, and so on. This model was considered conservative because it assumed that the average will be one-half the maximum available impact velocity. The LLNL analysis used the results of Fig. 3-11 (based on FRA data for 1976–82) and assumed distributions for impact angle and container orientation when impact occurs.

Although multiple factors affect the number of cars involved in a derailment, the SNL model was based only on speed as the independent variable. Three distributions were selected for the number of cars derailed as a function of speed. Two-car involvement distributions for collision accidents were postulated.

The results of the analysis are shown in Fig. 3-12 and Table 3-15, which show the upper and lower bounds from the various derailment and collision models. A limitation of these results is that the effect of having two or more cars containing the hazardous material of interest cannot be readily evaluated.

3.2.3.3 Crush Force

The basic approach used by SNL to determine the conditional probability of crush was to divide the potential crushing area of the actual railcars by an estimation of the area over which cars are scattered as a result of a derailment. Both of these values are a function of the speed at the time of derailment. A similar analysis was performed for collisions. The result of these analyses is that the conditional probability of crush is estimated as 2×10^{-3}, given a train accident.

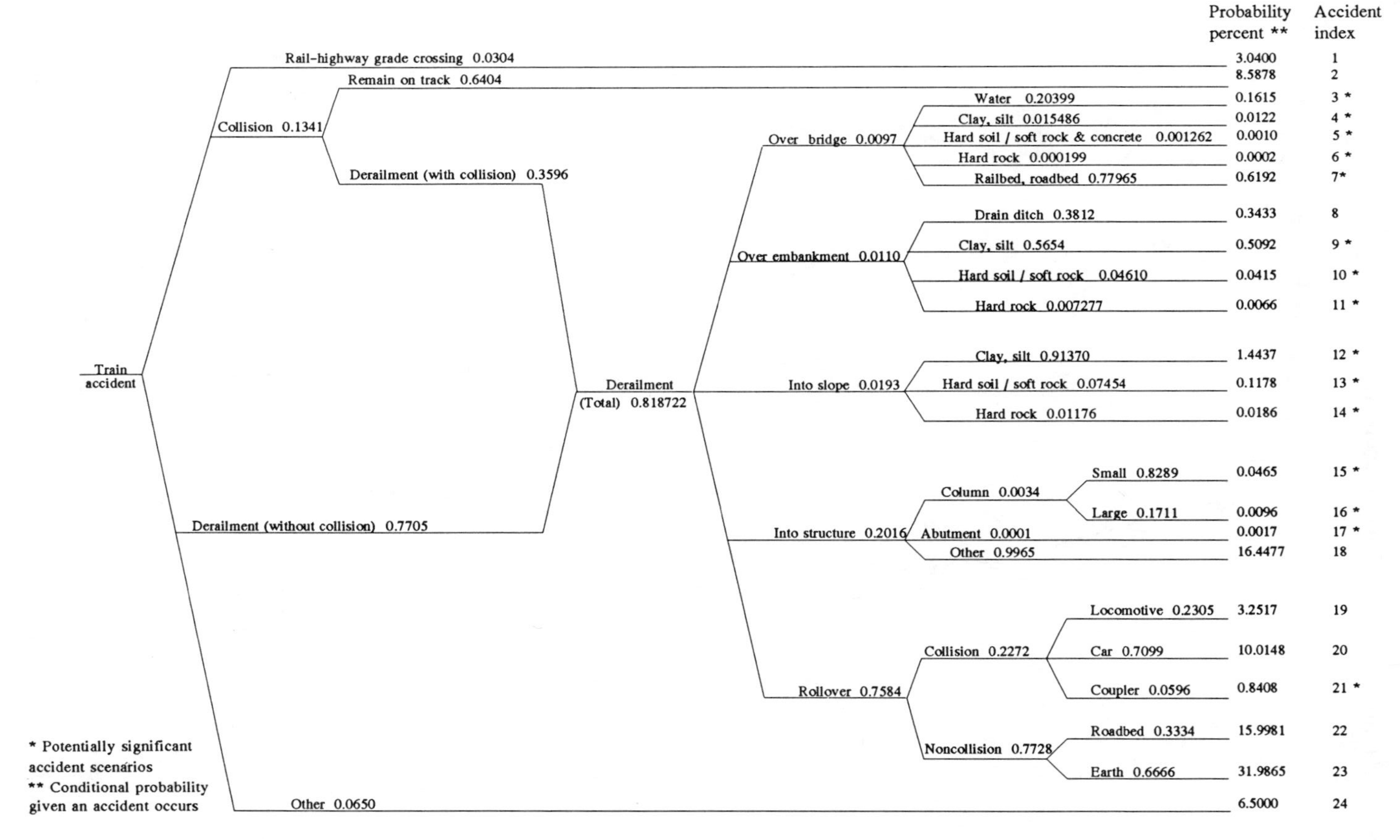

FIGURE 3-11. Train accident scenarios and their percent probabilities. Source: Fischer et al. 1987.

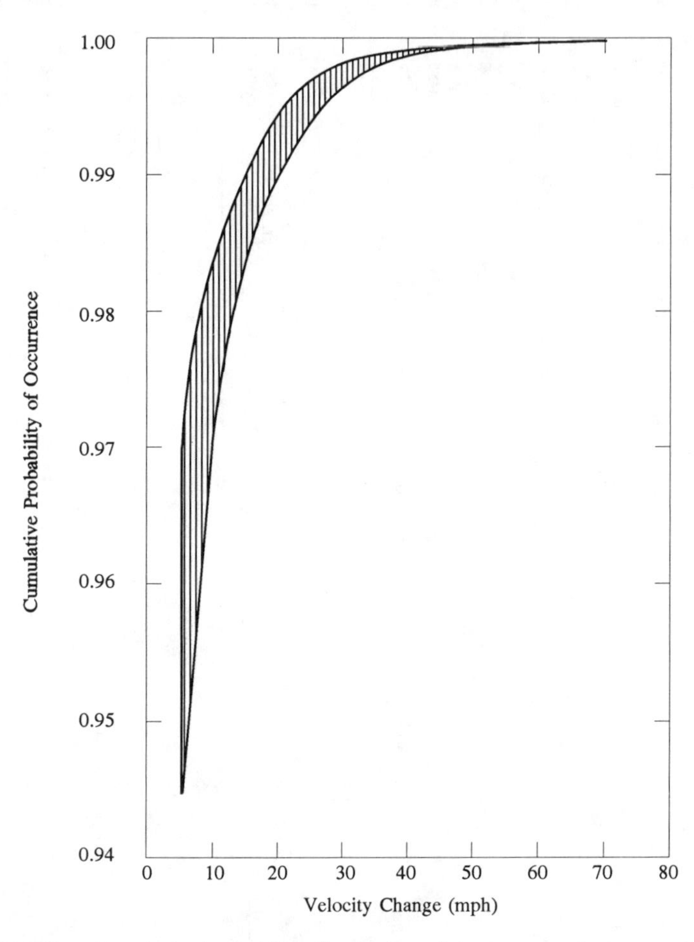

FIGURE 3-12. Cumulative probability distribution of the impact velocity magnitude for large containers in train accidents. Source: Dennis et al. 1978.

The accident severity was determined by using simple beam models with assumed probability distributions for the location of the container or the tank car underneath another car. The result is shown in Fig. 3-13.

3.2.3.4 Puncture Force

The number of tank car failures from puncture in a year, N, is:

$$N = (\text{number of car miles per yr})(\text{accident frequency per mile}) \times (\text{probability of a puncture environment})(\text{probability of puncture failure}) \quad (3\text{-}7)$$

TABLE 3-15. Impact velocity distribution for large containers in train accidents

Velocity change (mph)	Cumulative frequency of a velocity change less than or equal to the indicated velocity changes per train accident with one shipment car of interest per train	
	Case 1	Case 2
No impact	0.85077	0.91270
5	0.94615	0.96924
10	0.97102	0.98395
15	0.98301	0.99081
20	0.98990	0.99489
25	0.99386	0.99705
30	0.99631	0.99817
35	0.99780	0.99896
40	0.99878	0.99943
45	0.99935	0.99972
50	0.99970	0.99985
55	0.99983	0.99990
60	0.99990	0.99992
65	0.99995	0.99995
70	0.99997	0.99998
75	0.99999	1.00000
80	1.00000	

Source: Dennis et al. 1978.

For 1972 data, this is evaluated as:

$$20 = (3.16 \times 10^9)(1.0 \times 10^{-5})(\text{probability of a puncture force})(0.496) \quad (3\text{-}8)$$

where it is assumed that the typical tank car has a shell thickness of $7/16$ in., and 0.496 is the failure probability for that thickness as derived here. The conditional probability of a puncture force occurring, given an accident, is then $20/(3.16 \times 10^4)(0.496) = 1.28 \times 10^{-3}$. SNL decomposed this value into the product of the conditional probability of a collision or a derailment, given an accident, and the conditional probability of a puncture force occurring, given a collision or a derailment. For 1972, the former value is 0.90; thus the latter is $1.28 \times 10^{-3}/0.90 = 1.42 \times 10^{-3}$. SNL rounded the value to 1.5×10^{-3} for the final value of the conditional probability of a puncture force, given a collision or a derailment. This number could be readily recomputed on the basis of the RPI-ALR database described in Section 3.3.3. The data reported in the open literature are not complete, however.

The force magnitude model is based on the railcar coupler as the puncture probe and consists of determining the minimum railcar coupler velocity

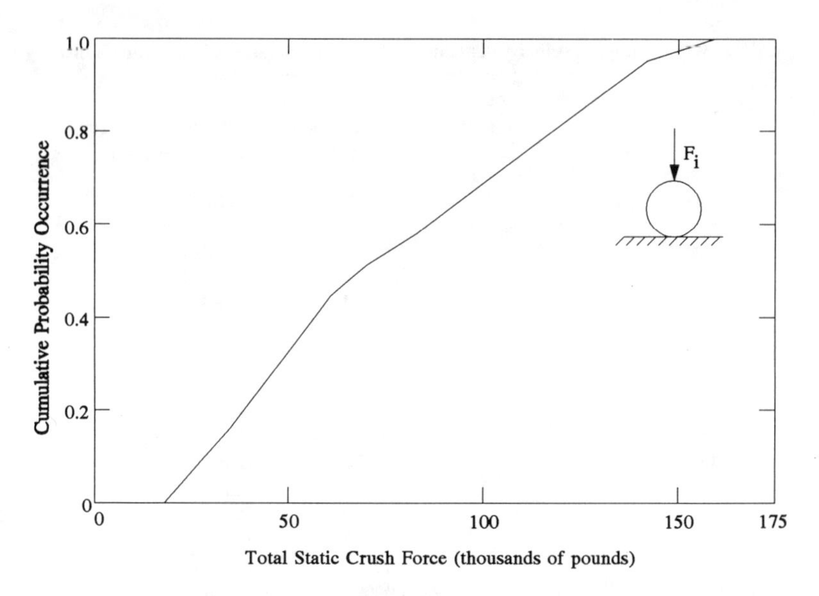

FIGURE 3-13. Cumulative probability distribution of total crush load for large containers in a train accident producing crush forces. Source: Dennis et al. 1978.

to puncture as a function of wall thickness. This velocity is combined with the probability distribution of the preaccident speed, which is assumed to be equal to the relative velocity between two cars. For thick-wall containers such as those used for spent nuclear fuel casks, the possibility of the coupler failing before the cask fails must also be considered. The results of these analyses are presented in Table 3-16. Note that the conditional probability of a puncture force, given an accident, has been included in the tabulated results.

3.2.4 Train Transport of Small Containers

Section 3.2.4 is based on Clarke et al. (1976) unless otherwise specified.

3.2.4.1 Fire Force

The small rail container fire data and duration model are identical to the large rail container fire model described in Section 3.2.3.1. The conditional probability of 0.184 car fires/train accident and Fig. 3-10 are applicable. The fire is considered external to the boxcar or other enclosure in which the small containers are transported. Alternatives for modeling the response of the container to the fire in order to use Fig. 3-10 are the same as for truck transport of small containers (Section 3.2.2.1).

TABLE 3-16. Probability of large container puncture during rail accident

Probability per reportable accident	Package wall thickness (in.)
7.41×10^{-4}	0.4375
6.90×10^{-4}	0.50
5.85×10^{-4}	0.75
4.90×10^{-4}	1.00
4.18×10^{-4}	1.25
3.37×10^{-4}	1.50
2.43×10^{-4}	1.75
1.52×10^{-4}	2.00
3.15×10^{-5}	2.50
4.70×10^{-6}	3.00
5.54×10^{-8}	4.00
9.14×10^{-12}	5.00

Source: Dennis et al. 1978.

3.2.4.2 Impact Force

The conditional probability of impact is the probability of collision or derailment, 0.90, as with large containers. The impact accidents and velocity distributions used for large containers also were used for small containers. An exception is that one-half of the velocity preceding derailment correlated better with the damage observed in this case, perhaps because a series of smaller impacts characterize the derailment environment. An analytical model similar to the truck impact model for small containers was used: a spring-mass model for the relative velocity between the boxcar walls and the containers, including the effects of coupler run-in. As was the case for the truck model, an overestimate of the velocity was obtained by assuming that a single container was originally located in the center of the boxcar. The results are presented in Fig. 3-14. The effect of converting the velocity into an equivalent drop height is shown in Fig. 3-15. The low magnitude of the impact threat from a conservative model justifies the selection of a simple (low calculational cost) method.

3.2.4.3 Crush Force

The static crush conditional probability, 7×10^{-5}, was found to be insignificant compared to dynamic crush, 0.90, and the analytical models for each were similar to those used for small container truck crush. The force magnitudes for static and dynamic crush were about the same. Therefore, static crush can be neglected compared with dynamic crush. The cumulative dynamic crush force probability distribution for a container in a collision or a derailment accident is presented in Fig. 3-16.

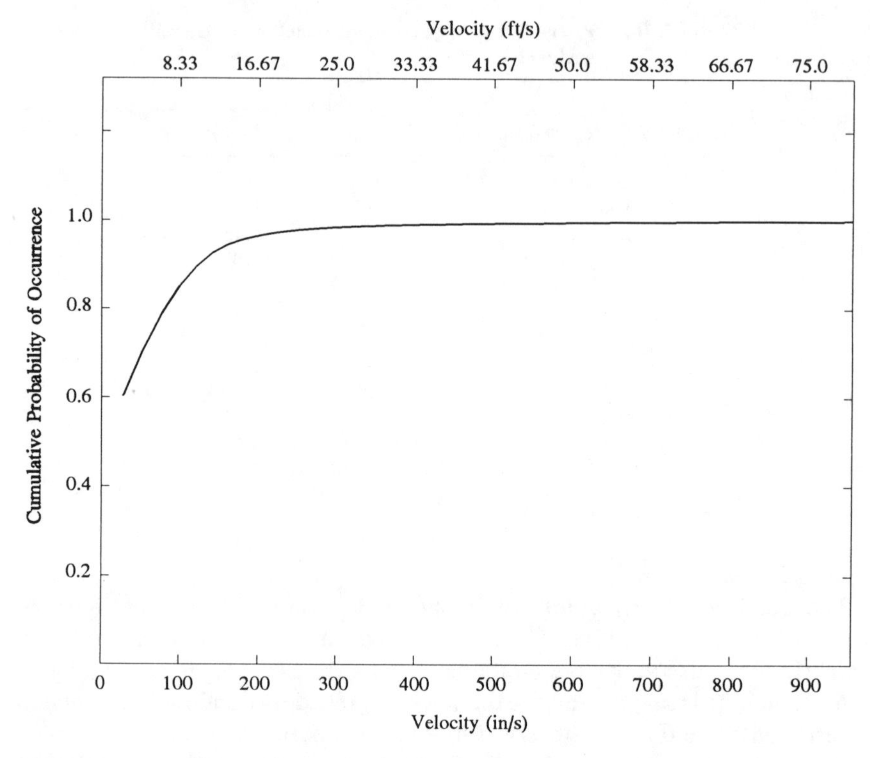

FIGURE 3-14. Cumulative probability distribution of maximum velocity of container against railcar wall, given a collision derailment accident. Source: Clarke et al. 1976.

3.2.4.4 Puncture Force

The puncture model is based on data for 1973 for accidents involving hazardous cargo transport by rail in small containers, which showed that 54.6% of the accidents producing damage (345 accidents) were due to puncture. Further, given an accident producing puncture in a container about the size of a 55-gal drum, an average of 2.3% of the drums were punctured. The conditional puncture probability is 0.90 collisions or derailments/train accident times 0.546 puncture accidents/cargo damage accident times 1 cargo damage accident/collision or derailment = 0.49 puncture accidents/train accident. The assumption that each collision or derailment produces cargo damage is conservative. If the number of containers of interest is less than 2.3%, then the conditional probability becomes 0.49 times 0.023 = 0.011 punctures/container/train accident.

The hard container puncture model was derived analogously to that for trucks described in Section 3.2.2.4, and the results are presented in Fig. 3-17. The bottom and dashed lines represent two different velocity assumptions.

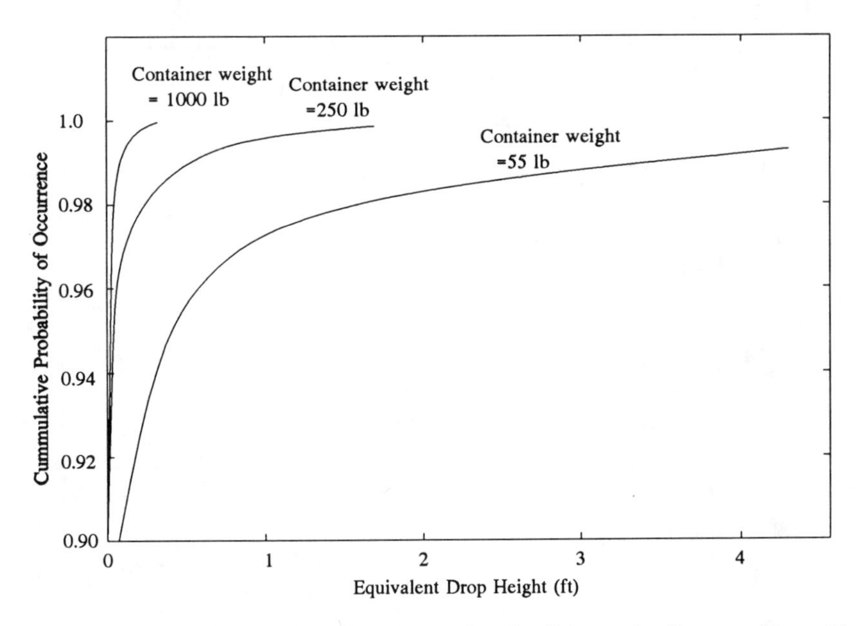

FIGURE 3-15. Cumulative probability distribution of collision or derailment accident with impact equivalent to given drop height. Source: Clarke et al. 1976.

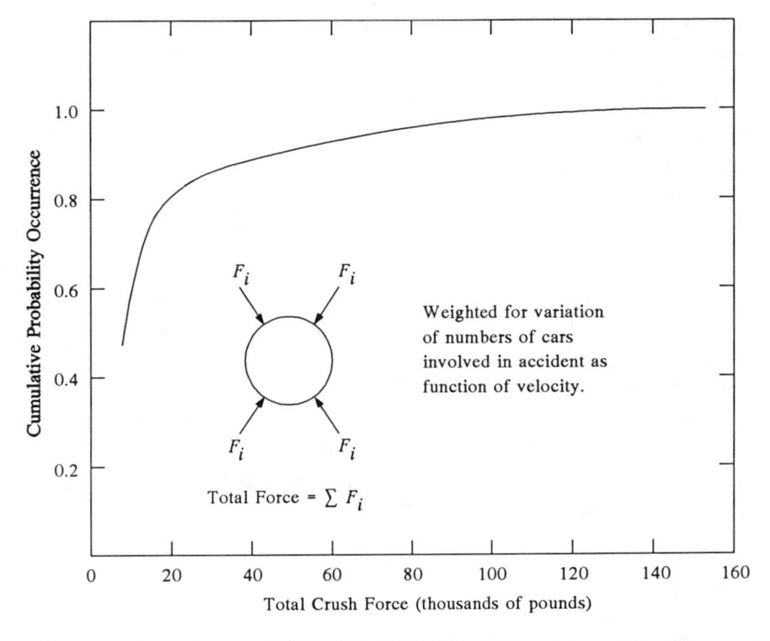

FIGURE 3-16. Cumulative probability distribution of total inertial crush load on a container in a train derailment or collision accident. Source: Clarke et al. 1976.

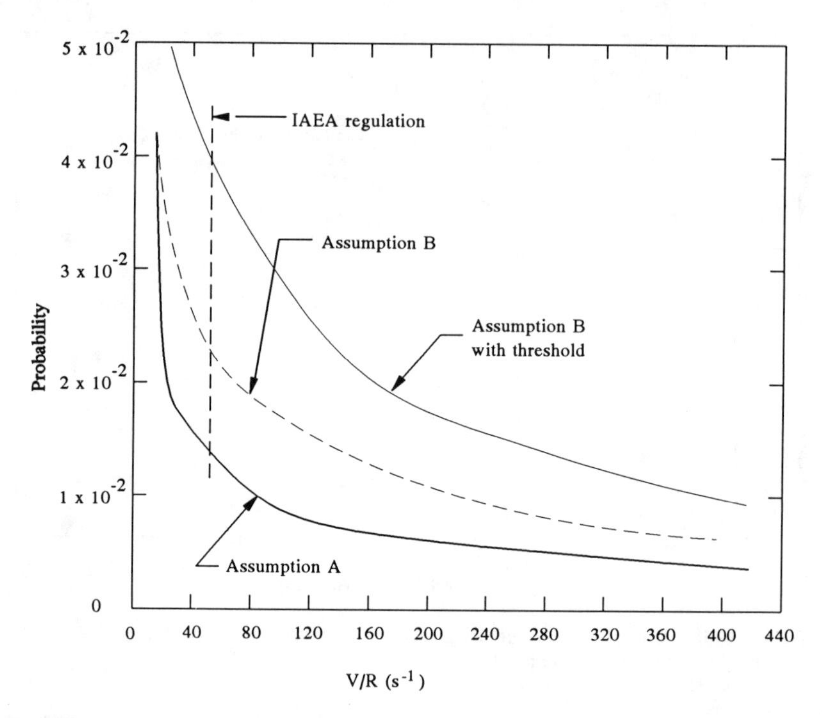

FIGURE 3-17. Probability of hard container experiencing puncture environment more severe than V/R, given that a soft container failed. Adapted from: Clarke et al. 1976.

The bottom solid line is based on the assumption that little movement (less than 6 in.) of the container occurs within the cargo area owing to close cargo spacing. The dashed line results when the derailment velocity is halved as described previously for the impact force. The probability for a constant V/R goes up because for a derailment velocity of v, the probability of a derailment at $v/2$ or greater is higher than the probability of a derailment at v or greater. The top line results when the assumption that any collision or derailment accident causes a puncture of general cargo is changed by assuming that a V/R threshold of 12.5 s^{-1} is required for any puncture to occur. The probability goes up for a constant V/R because the observed puncture rate, 54.6%, is now applicable to the portion of the distribution for which V/R exceeds 12.5 s^{-1}. The top line was considered to be an upper bound for the methodology used.

3.2.5 Summary

Clarke et al. (1976) and Dennis et al. (1978) combined data from the early 1970s with analytical engineering models to determine probabilistic distributions for the magnitude of accident forces. The engineering models were based either on a thick-walled bulk container or a load of small containers represented by 55-gal drums. The data used by Clarke et al. (1976) and Dennis et al. (1978) generally can be validated by comparison with truck and rail data through the early 1980s (Fischer et al. 1987) and truck data for the mid-1980s (Harwood and Russell 1990).

The details of the SNL analytical thermal model were released to Fischer et al. (1987) but are now considered proprietary. Fischer et al. (1987) adopted the SNL model. More work needs to be done to compute the probability that a specific railcar or railcars are involved in a train fire. Details for many SNL analytical models are not available for review or for recomputation with changed input parameters. The bulk rail impact model in particular needs to be reevaluated for multiple cars of interest in the same train. The bulk rail puncture model should be updated, but the needed data are proprietary to RPI-AAR. For either the bulk or small container rail models, the conditional probabilities of an accident force occurring are most accurate for impact and worst for puncture. The uncertainty is in the range of ± 10% to ± one order of magnitude, respectively. The force magnitude estimates have an uncertainty in the ± one order of magnitude range.

3.3 CONDITIONAL CONTAINER FAILURE PROBABILITY

The historical methodology (described in Section 2.4) relies on data described in this section.

3.3.1 Hazardous Materials Incident Reports

Any unintentional release of hazardous materials (except paint and paint-related material in packaging of 5 gal or less and batteries) during transportation, during loading or unloading, or in temporary storage related to transportation must be reported in writing to the DOT Research and Special Programs Administration (RSPA). Releases from any mode, excluding only bulk water transporters and motor carriers doing solely intrastate business, are included in a database called the Hazardous Material Information System (HMIS). The database is cataloged by 334 container types and 32 hazard class codes. The chemical name, amount released, whether a fire and/or an explosion occurred, vehicle speed, number of highway lanes, whether the highway was divided or undivided, and other information are available for each entry. Database searches can be obtained from the RSPA Office of Hazardous Materials Planning and Analysis (DHM-63), Washington, D.C. 20590.

An analysis by OTA (1986) resulted in the conclusion that HMIS underreporting of accidents may be as high as 50% for rail and 70% for highway. Harwood and Russell (1990) reexamined the OTA data and concluded that 130 accidents were common to HMIS and FHWA; HMIS had 152 unique accidents and FHWA had 138 unique accidents. Of the 420 accidents, 64% were in the FHWA database and 67% in the HMIS. Despite potential underreporting, OTA found the HMIS to be the best source available for container failure data. Table 3-17 (Harwood and Russell 1990) is based on HMIS data for hazardous materials (hazmat) and shows that although highway traffic accidents account for only 11% of reported incidents, they account for 35% to 68% of severe incidents, depending on the definition of severe. Valve or fitting failure is both the next most common source of incidents (24%) and the next highest contributor to severe incidents (10–29%).

Given a release, the HMIS contains a great deal of useful information about accidents and incidents, especially release amounts and container type. The HMIS cannot be used directly for the conditional probability of failure, given an accident, which is the primary focus of Section 3.3. The only readily available, systematic study of the HMIS that the author is aware of is for release amount by container type (Section 3.4).

3.3.2 Truck

The FHWA Office of Motor Carriers maintains a database of truck accident reports (see Table 3-1). An analysis of the database (Harwood and Russell 1990) is presented in Table 3-18. The average probability of a release, given an accident involving hazardous materials, was 14.0% for the years 1984 and 1985. The average for the years 1981 through 1985 was 15.2%; the rate seems fairly constant. Harwood and Russell (1990) acknowledge that the database may be biased owing to underreporting (see Section 3.3.1). Because accident reporting levels increase as accident severity increases, 15% release, given an accident, is considered by Harwood and Russell (1990) to be an upper bound. The data in Table 3-18 show that liquid tankers are slightly more likely than average to experience a cargo release, but hazardous materials in general freight, explosives, and bulk gases are less likely than average to experience a release. To the extent that the cargo type descriptions in Table 3-18 are representative of specific commodities (e.g., for LPG gases in bulk) the data can be used directly as the desired conditional release probability.

Harwood and Russell (1990) combined the results from Table 3-19 with the California, Illinois, and Michigan data (used to generate Table 3-4) and computed the results shown in Table 3-20. The data in Table 3-20 are for all trucks, and only minor variations were found for the corresponding values for single trailer combination trucks. Harwood and Russell (1990) acknowledge that in addition to the FHWA underreporting bias, the reporting

TABLE 3-17. Distribution of on-highway hazardous materials incidents by failure type and incident severity, 1981 through 1985

	Death only		Death or injury		Death, injury, or explosion		Death, injury, explosion, or fire		Death, injury, explosion, fire, or property damage over $100K		Death, injury, explosion, fire, or property damage over $50K		Death, injury, explosion, fire, or property damage over $10K		All reported incidents	
Failure type	No.	%	No.	%	No.	%	No.	%	No.	%	No.	%	No.	%	No.	%
Traffic accident	32	(91.4)	107	(35.5)	112	(34.7)	188	(41.7)	233	(46.4)	355	(56.1)	723	(68.1)	1,427	(10.8)
Body or tank failure	0	(0.0)	37	(12.3)	38	(11.8)	40	(8.9)	42	(8.4)	42	(6.6)	63	(5.9)	2,741	(20.2)
Valve or fitting failure	0	(0.0)	86	(28.6)	88	(27.2)	101	(22.4)	101	(20.1)	104	(16.4)	112	(10.5)	3,289	(24.3)
Cargo shifting	0	(0.0)	39	(13.0)	44	(13.6)	52	(11.5)	52	(10.4)	54	(8.5)	70	(6.6)	4,945	(36.5)
Fumes or venting	0	(0.0)	2	(0.7)	2	(0.6)	2	(0.4)	2	(0.4)	2	(0.3)	2	(0.2)	15	(0.1)
Other	3	(8.6)	30	(10.0)	39	(12.1)	68	(15.1)	72	(14.3)	76	(12.0)	92	(8.7)	1,100	(8.1)
TOTAL	35		301		323		451		502		633		1,062		13,547	

Source: Harwood and Russell 1990.

TABLE 3-18. Distribution of FHWA-reported truck accidents by cargo type, 1984 through 1985

	Accidents Involving Trucks Not Carrying Hazmat		Accidents Involving Trucks Carrying Hazmat						
			Combined		No release		Hazmat release		
Cargo type	No.	%	No.	%	No.	%	No.	%	Release probability (%)
General freight	23,651	(33.7)	741	(20.1)	680	(21.4)	61	(11.8)	8.2
Gases in bulk	42	(0.1)	259	(7.0)	238	(7.5)	21	(4.1)	8.1
Solids in bulk	1,310	(1.9)	40	(1.1)	28	(0.9)	12	(2.3)	30.0
Liquids in bulk	1,618	(2.3)	1,831	(49.6)	1,486	(46.8)	345	(66.6)	18.8
Explosives	12	(0.1)	70	(1.9)	63	(2.0)	7	(1.4)	10.0
Empty	15,989	(22.8)	220	(6.0)	210	(6.6)	10	(1.9)	4.5
Other	27,478	(39.2)	529	(14.3)	467	(14.7)	62	(12.0)	11.7
TOTAL	70,100		3,690		3,172		518		14.0

Source: Harwood and Russell 1990.

TABLE 3-19. Probability of release, given that an accident has occurred, as a function of accident type

Accident type	Probability of release
Single-Vehicle Noncollision Accidents	
Run-off-road	0.331
Overturned (in road)	0.375
Other noncollision	0.169
Single-Vehicle Collision Accidents	
Collision with parked vehicle	0.031
Collision with train	0.455
Collision with nonmotorist	0.015
Collision with fixed object	0.012
Other collision	0.059
Multiple-Vehicle Collision Accidents	
Collision with passenger car	0.035
Collision with truck	0.094
Collision with other vehicle	0.037

Source: Harwood and Russell 1990.

TABLE 3-20. Probability of hazmat release, given that an accident has occurred

Highway Class		Probability of Hazmat Release, Given an Accident			
Area type	Roadway type	California	Illinois	Michigan	Weighted average[1]
Rural	Two-lane	0.100	0.074	0.073	0.086
Rural	Multilane undivided	0.100	0.071	0.064	0.081
Rural	Multilane divided	0.087	0.064	0.062	0.082
Rural	Freeway	0.083	0.111	0.095	0.090
Urban	Two-lane	0.077	0.059	0.069	0.069
Urban	Multilane undivided	0.064	0.052	0.055	0.055
Urban	Multilane divided	0.068	0.048	0.058	0.062
Urban	One-way street	0.066	0.050	0.056	0.056
Urban	Freeway	0.062	0.055	0.067	0.062

Source: Harwood and Russell 1990.
[1]Weighted by veh-mi of truck travel.

threshold for FHWA of $4,400 is 5 to 10 times higher than those of the three states used to generate Table 3-20; however, the available data are insufficient to compute an adjustment.

3.3.3 Train

The BOE has maintained a computerized database on railway incidents and accidents since 1975. The sources of these data are the HMIR reports and

TABLE 3-21. Leakage frequency by commodity hazard class

Commodity/ hazard class[1]	1986			1987			1988			1989			1990			1991		
	Car leaks	Cars shipd[2]	Rate[3]	Car leaks	Cars shipd	Rate	Car leaks	Cars shipd	Rate	Car leaks	Cars shipd	Rate	Car leaks	Cars shipd	Rate	Car leaks	Cars shipd	Rate
Ammonia	82	43	1.91	94	54	1.74	67	54	1.24	62	67	0.93	68	56	1.21	59	71	0.83
Nonflammable gas (Division 2.2)	137	102	1.34	156	126	1.24	135	125	1.08	151	152	0.99	139	134	1.04	107	161	0.66
LPG	91	111	0.82	69	127	0.54	61	134	0.46	71	175	0.41	133	155	0.85	64	222	0.29
Flammable gas (Division 2.1)	101	136	0.74	80	155	0.52	73	164	0.45	78	215	0.36	142	190	0.75	75	266	0.28
Flammable liquid (Class 3)	229	157	1.46	190	185	1.03	237	188	1.26	247	214	1.15	246	211	1.17	189	254	0.74
Combustible liquid	47	74	0.64	50	68	0.74	74	62	1.19	82	91	0.9	74	90	0.82	74	114	0.65
Poison A (Division 2.3)	0	1	0	0	1	0	0	1	—	0	2	—	0	2	—	0	11	—
Poison B (Division 6.1)	12	12	1	7	7	1	8	12	0.67	1	15	0.07	7	14	0.5	7	17	0.41
Sulfuric acid	123	54	2.28	81	51	1.59	110	51	2.16	118	71	1.66	148	64	2.31	116	80	1.45
Phosphoric acid	47	28	1.68	45	30	1.5	26	27	0.96	66	32	2.06	75	31	2.42	65	35	1.86
Hydrochloric acid	61	11	5.55	58	12	4.83	65	11	5.91	172	15	11.5	173	14	12.4	108	20	5.40
Corrosive matl (Class 8)	387	203	1.91	376	229	1.64	394	223	1.77	532	273	1.95	575	253	2.27	449	311	1.44
Hazardous waste	No info available for these years.						3	2	0.15	10	3	0.33	14	4	0.35	13	6	0.22
Others	27	51	0.53	55	74	0.74	61	81	0.75	64	129	0.5	61	122	0.5	61	179	0.34
TOTAL	940	736	1.28	914	845	1.08	985	857	1.15	1165	1092	1.07	1258	1018	1.24	975	1308	0.75

Source: BOE 1992a.

[1]Selected commodities are included in hazard class totals. TOTAL includes other hazard classes.

[2]"Cars Shipd" is loaded tank car originations (×1000) as reported to TRAIN II.

[3]"Rate" is number of leaking incidents divided by "Cars Shipd." (Leaks per 1,000 cars shipped.)

on-scene reports from BOE inspectors. The nonaccident release portion is reported annually for tank cars (BOE 1992b). The cause of release (e.g., vapor line open) is reported for each commodity. The number of originations of tank cars of that commodity also is reported; thus, a release frequency on a per trip basis is available directly. No indication of leak size is provided. A summary of leak frequency by hazard class is presented in Table 3-21 (BOE 1992a). The commodities responsible for a majority of leaks in a hazard class are listed individually. Generally, the most transported commodities (e.g., LPG) are responsible for the most leaks. An exception is chlorine, which consistently has a leak rate of approximately 1 per 10,000 cars shipped: 0.5 per 10,000 in 1991 and 1.3 per 10,000 in 1990 (BOE 1992a). According to the BOE, this exception may be due to the special design features of the tank car and the standardized valving arrangement. Corrosive materials are responsible for 46% of the leaks but only 24% of the tank car originations. These cars generally are equipped with rupture disks rather than spring-loaded safety relief valves because of the corrosive nature of the cargo. Safety valves are designed to reseal after functioning, but rupture disks are not. Rupture disks are vulnerable to improper installation and pressure spikes due to movement of tank cars in rail yards (BOE 1992a). The BOE is unable to explain the low leak rate of flammable gas compared to nonflammable gas; both are shipped in the same car types with the same valving arrangements. A summary of the sources of leaks for the years 1986 through 1991 is presented in Table 3-22. Note that the number of sources of leaks is usually greater than the number of leaky tank cars owing to multiple leaks.

The FRA, Office of Safety, publishes an annual bulletin with accident/incident data (FRA 1992). Table 3-23 shows accident data for the years 1986 through 1991, and Table 3-24 shows a further breakdown for 1991. No information is reported on the size of the release or the specific failure mode. FRA does report the cause of the accident (e.g., roadbed defect) and the accident breakdown information of Table 3-23 by state and by railroad.

The Railway Progress Institute (RPI) and the AAR have maintained a database on tank cars that have been damaged in accidents since 1965. Damage and spillage from normal operation and maintenance are not included. Tank cars with damage limited to items common to all freight cars such as trucks or brake equipment also are not included (Davis and Fowler 1990). The 22 years of data for the time span from 1965 through 1986 have been examined extensively. Fig. 3-18 shows an overview of the tank car damage that occurred as well as those that lost lading and the cause of the loss (Phillips 1990). The category "no damage (loose)" means that the car had damage other than at the top fitting, the car overturned to some degree, and product was released through a loose closure or one that was not gasketed

TABLE 3-22. Sources of nonaccident releases, 1986 through 1991

Year	1986	1987	1988	1989	1990	1991
Sources of Leaks						
Safety device	327	327	352	499	563	441
(Valve functioned)	(51)	(63)	(77)	(91)	(79)	(56)
(Disk ruptured)	(157)	(165)	(206)	(336)	(388)	(334)
Bottom fittings	183	138	174	172	152	139
Manway	297	267	280	323	362	253
Liquid line	143	112	106	126	141	97
Other top fittings	86	116	108	120	144	98
Shell or head	30	32	50	39	34	21
Total Leaking Tank Cars[1]	940	914	985	1165	1257	975

Source BOE 1992a.
[1]These numbers denote individual leaks; this includes multiple leaks from the same car.

TABLE 3-23. Train accidents involving hazardous materials

Year	Total number of accidents	Accidents in which a hazmat car was damaged or derailed	Accidents in which there was a release of hazardous materials	Accidents that resulted in evacuation
1981	586	353	77	27
1982	494	286	59	13
1983	422	240	52	16
1984	436	237	54	17
1985	415	245	54	22
1986	364	185	51	32
1987	352	186	50	28
1988	475	237	44	32
1989	517	251	55	28
1990	466	236	35	20
1991	525	293	47	28

Source: FRA 1992 and FRA 1987.

properly. Similar information for 1980 to 1986 is presented in Table 3-25 for a commodity/tank car grouping as shown in Table 3-26 (Davis and Fowler 1990). The number of loaded tank car shipments and the lading loss failures caused by all punctures and by punctures under 8 in. also are shown in Table 3-25. The values when aggregated for pressure cars, nonpressure cars, and all tank cars are shown in Table 3-27. Also shown are four conditional probabilities defined in the table footnotes as rates 1 through 4.

The conditional probability of damage, given a shipment (rate 1), varies

TABLE 3-24. Train accidents involving hazardous materials, 1991

	Cars						
Type of accident	Consists carrying	In consist	Containing hazmat	Damaged w/hazmat	Releasing hazmat	People evacuated	Damage to equipment
Derailment	396	27,111	3,221	634	71	6,017	35,394,334
Head-on collision	5	397	123	10	5	12	4,118,112
Rear-end collision	2	222	40	1	—	—	35,120
Side collision	47	2,282	316	40	4	—	1,811,317
Raking collision	10	314	29	8	—	—	3,629,937
Broken train collision	2	297	61	—	—	—	123,000
Hwy-rail crossing	19	1,424	102	6	1	230	1,669,776
RR grade crossing	—	—	—	—	—	—	—
Obstruction	7	443	27	—	—	—	413,510
Explosion-detonation	—	—	—	—	—	—	—
Fire or violent rupture	2	187	12	—	—	—	19,200
Other	55	2,351	234	52	3	200	2,117,987
TOTAL	545	35,028	4,165	751	84	6,459	49,332,293

Source: FRA 1992.

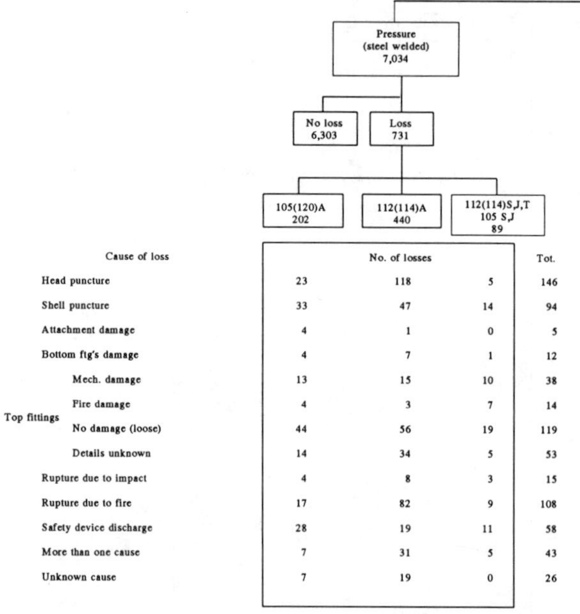

Cause of loss		105(120)A	112(114)A	112(114)S,J,T 105 S,J	Tot.
		No. of losses			
Head puncture		23	118	5	146
Shell puncture		33	47	14	94
Attachment damage		4	1	0	5
Bottom ftg's damage		4	7	1	12
Top fittings	Mech. damage	13	15	10	38
	Fire damage	4	3	7	14
	No damage (loose)	44	56	19	119
	Details unknown	14	34	5	53
Rupture due to impact		4	8	3	15
Rupture due to fire		17	82	9	108
Safety device discharge		28	19	11	58
More than one cause		7	31	5	43
Unknown cause		7	19	0	26

a) All tank cars, U.S. and Canada, hazardous and nonhazardous commodities
b) There were 135 nonpressure S/S cars and 128 nonpressure U/F cars, all with no loss, but unknown whether insulated or noninsulated.

FIGURE 3-18. Tank cars damaged in accidents, 1965 through 1986 overview. Source: Phillips 1990.

dramatically from the top time period (1965-79) to the bottom time period (1980-86) for pressure cars but less so for nonpressure cars. This effect can be attributed to a lowering of the overall accident rate and to the introduction of head and thermal shields and shelf couplers. The introduction of shelf couplers for 112 and 114 cars began in 1978, and head and thermal shields began in 1979. The effect on head punctures alone for 112 and 114 cars for the shelf coupler and head shield combination was determined to be 91% reduction exclusive of the 44% accident rate reduction, for a combined reduction of 95% (Phillips and Role 1989). Additional safety improvements on certain nonpressure and pressure cars continued through the data period. Because the safety improvements span both tables, the use of 1980 to 1986 data should be conservative for current shipments. Note that the conditional probability of lading loss, given damage (rate 2), does not change significantly between the 1965 to 1979 and the 1980 to 1986 time frames. The lower

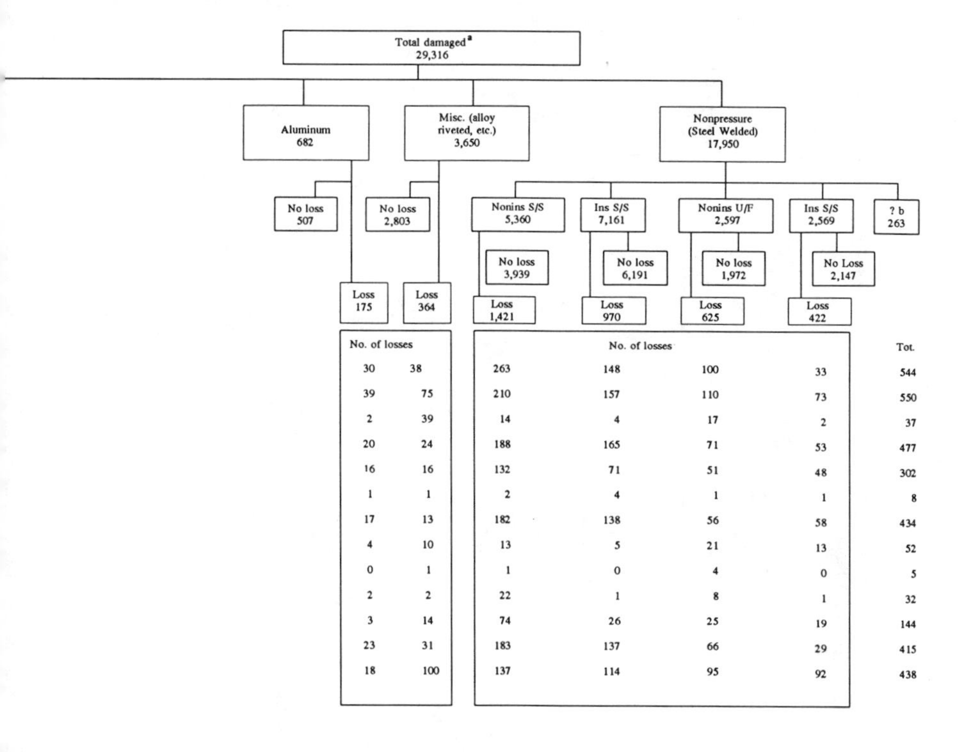

FIGURE 3-18. *Continued*

line of Fig. 3-19 shows the yearly variation of the conditional probability of release, given damage, for hazardous materials (Harvey et al. 1987). The top line in Fig. 3-19 is the conditional probability of release from any damaged hazardous material car, given that at least one car carrying hazardous materials has been damaged.

3.3.4 Summary

Highway data through the mid-1980s were presented by Harwood and Russell (1990), aggregated by either failure type, cargo type, accident type, or highway type. Railway tank car data through 1991 were reported by the AAR-BOE for nonaccident releases by commodity and commodity class. Railway tank car accident data through 1986 were reported by the FRA for hazardous materials as a group, by Phillips (1990) for rail tank car types and

TABLE 3-25. Number of tank car shipments of hazardous materials and tank cars damaged and with lost lading of hazardous materials in railroad accidents in the United States and Canada, 1980 through 1986

Tank car group and hazardous material class as defined in Table 3-26	Number of loaded shipments in tank cars (thousands)	Number of damaged tank cars (thousands)	Number of tank cars with lost lading (thousands)	Number[1] of tank cars with puncture (thousands)	Number[2] of tank cars with puncture under 8 in. (thousands)
Nonpressurized Tank Cars					
1	1,250	1.96	0.374	0.109	0.051
2	1,200	0.646	0.123	0.032	0.011
3	780	0.315	0.060	0.018	0.009
4	1,320	1.35	0.258	0.092	0.020
5	113	0.084	0.016	0.004	0.004
6	52.5	0.021	0.004	0.002	0.002
7	26.3	0.121	0.023	0.012	0.0
8	31.9	0.0369[1]	0.010	0.002	0.0
TOTAL	4,773.7	4.5339[1]	0.868	0.271	0.097
Pressurized Tank Cars					
9	1,410	0.678	0.071	0.022	0.016
10	477	0.124	0.013	0.003	0.0
11	605	0.162	0.017	0.004	0.002
12	37.5	0.00955[1]	0.001	0.0	0.0
TOTAL	2,529.5	0.97355[1]	0.102	0.029	0.018
TOTAL ALL TANK CARS	7,303.2	5.50745[1]	0.970	0.300	0.115

Source: Davis and Fowler 1990.
[1]The fractional numbers are the result of interpolation.
[2]Unknowns are prorated.

TABLE 3-26. Categorization of hazardous material classes and tank car groups

Class number	Class description
1	Flammable liquids in nonpressurized noninsulated tank cars [groups 1—riveted, 2—103W(203W), 4—underframe 111A(211A), 6—stub sill 111A(211A)].
2	Combustible liquids in nonpressurized noninsulated tank cars [groups 1—riveted, 2—103W(203W), 4—underframe 111A(211A), 6—stub sill 111A(211A)].
3	Corrosive material (NaOH) in nonpressurized insulated tank cars [groups 3—103W(104W), 5—underframe 111A(211A), 7—stub sill 111A(211A)].
4	Corrosive materials (except NaOH) in nonpressurized noninsulated tank cars [1—riveted, 2—103W(203W), 4—underframe 111A(211A), 6—stub sill 111A(211A)].
5	Poison B (other than MFAK) in nonpressurized noninsulated tank cars [groups 1—riveted, 2—103W(203W), 4—underframe 111A(211A), 6—stub sill 111A(211A)].
6	Flammable solids in nonpressurized insulated tank cars [groups 3—103W(104W), 5—underframe 111A(211A), 7—stub sill 111A(211A)].
7	Other regulated materials in nonpressurized noninsulated tank cars [groups 1—riveted, 2—103W(203W), 4—underframe 111A(211A), 6—stub sill 111A(211A)].
8	Oxidizer materials in aluminum tank cars [group 11—miscellaneous 113A, 204, 115A(206), etc.].
9	Flammable gases in pressurized noninsulated tank cars [group 9—112A(114A) during the 1965-79 period; group 10—105S, J and 112(114)S, J, T with thermal shields, head shields, and shelf couplers during the 1980-86 period].
10	Nonflammable gases (NH_3) in pressurized noninsulated tank cars [group 9—112A(114A) during the 1965-79 period; group 10—105S, J and 112(114)S, J, T with thermal shields, head shields, and shelf couplers during the 1980-86 period].
11	Nonflammable gases (except NH_3) in pressurized insulated tank cars [group 8—105A(120A)].
12	Poison B (MFAK) in pressurized insulated tank cars [group 8—105A(120A)].

Source: Davis and Fowler 1990.

cause of release, and by Davis and Fowler (1990) for tank car group and hazardous material class. The uncertainty is probably about a factor of two for the average of the aggregated group reported. Owing to safety improvements, the tank car failure data are generally conservative. For a particular accident involving a particular hazardous material, the uncertainty is in an order of magnitude range.

The historical and predictive approaches to transportation risk methodology are described in Section 2.4. The historical approach relies on data of the type presented in this section, and the predictive approach to container failure relies on analyses described in Chapter 5.

TABLE 3-27. Summary of tank car hazardous material shipments and railroad accident loss rates in the United States and Canada, 1965 through 1979 compared to 1980 through 1986

	Number of loaded shipments in tank cars (thousands)	Number of damaged tank cars (thousands)	Number of tank cars with lost lading (thousands)	Number of tank cars with puncture (thousands)	Number of tank cars with puncture under 8 in. (thousands)
Tank Cars 1965 through 1979					
Nonpressurized Tank Cars					
Total number of tank cars	7,385	11.791	2.251	0.837	0.232
Average number of cars per yr	492.333	0.786	0.150	0.056	0.015
Rates 1	1.0	0.001 597	0.000 305	0.000 113	0.000 031
2		1.0	0.190 9	0.071 0	0.019 7
3			1.0	0.372	0.103
4				1.0	0.277
Pressurized Tank Cars					
Total number of tank cars	3,912.5	5.846 4[a]	0.612	0.236	0.080
Average number of cars per yr	260.833	0.390	0.041	0.016	0.005
Rates 1	1.0	0.001 494	0.000 156	0.000 060	0.000 020
2		1.0	0.104 7	0.040 4	0.013 7
3			1.0	0.386	0.131
4				1.0	0.339
All Tank Cars					
Total number of tank cars	11,297.5	17.637 4[a]	2.863	1.073	0.312
Average number of cars per yr	753.167	1.176	0.191	0.072	0.021
Rates 1	1.0	0.001 561	0.000 253	0.000 095	0.000 028
2		1.0	0.162 3	0.060 8	0.017 7
3			1.0	0.375	0.109
4				1.0	0.291

	Tank Cars 1980 through 1986				
Nonpressurized Tank Cars					
Total number of tank cars	4,773.7	4.533 9[a]	0.868	0.271	0.097
Average number of cars per yr	681.957	0.648	0.124	0.039	0.014
Rates 1	1.0	0.000 950	0.000 182	0.000 057	0.000 020
2		1.0	0.191 4	0.059 8	0.021 4
3			1.0	0.312	0.112
4				1.0	0.348
Pressurized Tank Cars					
Total number of tank cars	2,529.5	0.973 55[a]	0.102	0.029	0.018
Average number of cars per yr	361.357	0.139	0.015	0.004	0.003
Rates 1	1.0	0.000 385	0.000 040	0.000 011	0.000 007
2		1.0	0.104 8	0.029 8	0.018 5
3			1.0	0.284	0.176
4				1.0	0.621
All Tank Cars					
Total number of tank cars	7,303.2	5.507 45[a]	0.970	0.300	0.115
Average number of cars per yr	1,043.314	0.787	0.139	0.043	0.016
Rates 1	1.0	0.000 754	0.000 133	0.000 041	0.000 016
2		1.0	0.176 1	0.054 5	0.020 9
3			1.0	0.039	0.119
4				1.0	0.383

Source: Davis and Fowler 1990.

Notes: Rate 1 = number of affected tank cars per loaded tank car shipment. Rate 2 = number of affected tank cars per accident damaged tank car. Rate 3 = number of affected tank cars per damaged tank car with lost lading. Rate 4 = number of affected tank cars per punctured tank car. [a]The fractional numbers are the result of interpolation.

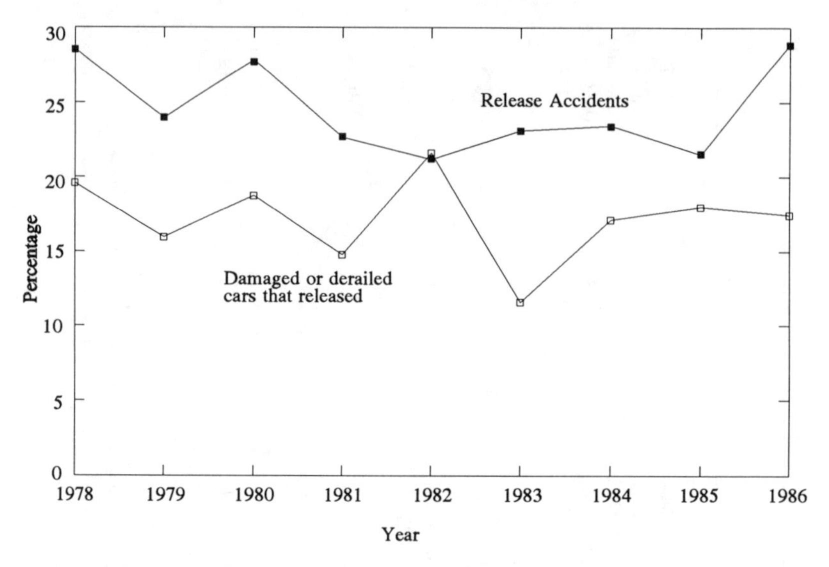

FIGURE 3-19. Percentage of damaged or derailed hazardous materials cars that released and the percentage of accidents with damaged or derailed hazardous materials cars that were release accidents, 1978 through 1986. Source: Harvey et al. 1987.

3.4 RELEASE AMOUNT

The most common approach to using the data in either Section 3.2 or Section 3.3 is to estimate conditional failure probabilities for a spectrum of release sizes. The smallest release size usually is associated with nonaccident incident leaks, and the largest is associated with gross failure of the container. Intermediate release sizes usually are associated with specific features (e.g., shearing a line of a certain size) or with failures producing holes in the container, in the 0.5- to 2-in. diameter range. The total number of postulated failure types depends on the objectives of the analysis. Use of one, two, or three mechanical failure types is most common although four or more sometimes are used. A relief valve lifting and gross rupture from overpressure are the usual failure modes from thermal events.

For each failure type, a release amount is needed for consequence analysis. In the general case, given a failure type, there is a range of release amounts, each associated with a particular probability of occurrence. For gross failure, the release amount will be nearly complete every time. Other release amounts depend on where the hole is relative to a liquid level, whether the liquid is under pressure, whether thermal forces are present, and so on. Many of these release amount situations lend themselves to analytical solutions of mass, energy, and momentum equations—a potentially expensive cost to the risk analysis project. Analytical determination of the release amount is covered in Chapter 6. The purpose of this chapter is to present commonly available analyses of the release amount data and to indicate where actual data can be found for more rigorous evaluations.

3.4.1 Hazardous Material Incident System

Abkowitz et al. (1984) used the HMIS to estimate the fractional release amount by container type for highway transport. The container types in the database were aggregated into eight container types, including cylinders, tanks, and fiber boxes. Releases from accidents and from nonaccident incidents were combined into a constant term and a term dependent on the number of shipment miles. The fraction released at terminal points also was estimated. An exposure term was computed by estimating the average shipment volumes, the average number of miles per shipment, the accident rate for various highway types, and the percentage of accidents that lead to a release from 1980, 1981, and/or 1982 data from five states. According to Harwood and Russell (1990), the assumed 20% of accidents that lead to a release is the weakest element in the Abkowitz et al. (1984) model.

The results of Abkowitz et al. (1984) are expressed in fraction leaked per mile. For the risk analysis approaches discussed in this book, the desired parameter is the amount leaked, given a leak occurs. An updated calculation that could distinguish between accident releases and various types of non-accident releases, as well as various types of loading and unloading accidents, would be very useful. To the extent possible different types of accident failure modes should be distinguished; for example, a fire failure release is different from a puncture failure release. In addition, to the extent possible a range of release amounts, each associated with a probability of occurrence, should be determined.

3.4.2 Cumulative Probability Distribution

A particularly useful way to present the data is in a cumulative probability distribution, as shown in Fig. 3-20 for gasoline trucks (Stewart and van Aerde 1990). In this case the data are presented not for container failure type but for accident type. Only 26 accident events were included in the analysis; therefore, the statistical significance is of concern. Work on gasoline is continuing, and a curve for LPG also is being developed at the University of Waterloo (personal communication with F. F. Sacomanno, April 1992).

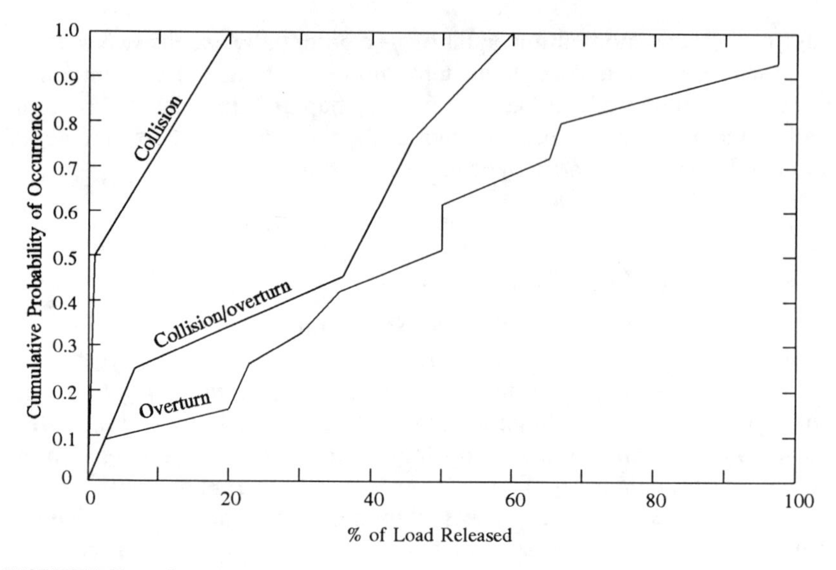

FIGURE 3-20. Cumulative probability distribution of percentage spilled for tanker-tractor carrying gasoline (following overturn, collision/overturn, and collision accidents). Source: Stewart and Van Aerde 1990.

3.4.3 Train

The RPI-AAR database described in Section 3.3.3 has been analyzed for nonaccident and accident release amounts. Release amounts in Table 3-28 (Davis and Fowler 1990) correspond to the failure probabilities in Table 3-25; therefore, average release amounts can be determined for three release types for 12 commodity/tank car groups. The data shown in Table 3-29 (Davis and Fowler 1990) correspond to the failure probabilities in Table 3-27. The average spill amounts are given directly for broad classes of tank cars. The total gallons lost per year for incidents and accidents are given in Fig. 3-21 (Harvey et al. 1987). Using this figure in conjunction with Fig. 3-19, one can conclude that although the conditional probability of release, given that damage occurred, varied from 12% to 22%, the total amount of the annual releases varied from about 400,000 gal to 800,000 gal in 1980 through 1986. From Fig. 3-19, the number of cars releasing hazardous materials was higher in 1982, when the annual release amount was low. The data indicate that the amount released varies substantially for each accident.

3.4.4 Summary

The HMIS offers great potential for providing release amount, given either a nonaccident or an accident release. The release amount could be deter-

TABLE 3-28. Number of gallons of hazardous material shipped in tank cars and lost from damaged tank cars in railroad accidents in the United States and Canada, 1980 through 1986

Tank car group and hazardous material class as defined in Table 3-26	Total gallons shipped (millions)	Gallons in damaged tank cars (millions)	Gallons lost from damaged tank cars (millions)	Gallons lost from punctured tank cars[1] (millions)	Gallons lost from tank cars w/ 8 in. or less punctures[1] (millions)
Nonpressurized Tank Cars					
1	28,700	44.9	3.05	1.490	0.704
2	25,800	13.9	0.877	0.458	0.187
3	12,100	4.88	0.227	0.074 3	0.002 7
4	20,200	20.80	1.62	0.824	0.139
5	2,280	1.70	0.045 9	0.041 400	0.000 004
6	845	0.338	0.012 5	0.000 020	0.000 020
7	505	2.32	0.230	0.174	0.0
8	383	0.443	0.025 4	0.013	0.0
TOTAL	90,813	89.281	6.087 8	3.074 72	1.032 724
Pressurized Tank Cars					
9	42,600	20.5	0.838	0.294	0.165
10	16,000	4.16	0.078 2	0.052 3	0.0
11	10,300	2.76	0.112	0.040 5	0.034
12	228	0.058	0.006	0.0	0.0
TOTAL	69,128	27.478	1.034 2	0.386 8	0.199
TOTAL ALL TANK CARS	159,941	116.759	7.122	3.461 52	1.231 724

Source: Davis and Fowler 1990.
[1]Unknowns prorated.

TABLE 3-29. Summary of gallons of hazardous material shipped in tank cars and railroad accident loss rates in the United States and Canada, 1965 through 1979 compared to 1980 through 1986

	Total gallons shipped	Gallons in damaged tank cars	Gallons lost from damaged tank cars	Gallons lost from punctured tank cars	Gallons lost from tank cars w/ 8 in. or less punctures
Tank Cars 1965 through 1979					
Nonpressurized Tank Cars					
Total gallons (millions)	117,461.	198.110	18.350	9.264 8	1.873 2
Average gallons per yr (millions)	7,830.733	13.207	1.223	0.618	0.125
Gallons per tank cars	15,905.	16,802.	8,152.	11,069.	8,074.
Rates 1	1.0	0.001 687	0.000 156	0.000 079	0.000 016
2		1.0	0.092 6	0.046 8	0.009 5
3			1.0	0.505	0.012
4				1.0	0.202
Pressurized Tank Cars					
Total gallons (millions)	100,122.	165.891	10.206 5	5.210 102	1.372
Average gallons per yr (millions)	6,674.800	11.059	0.680	0.347	0.091
Gallons per tank car	25,590.	28,375.	16,677.	22,077.	17,150.
Rates 1	1.0	0.001 657	0.000 102	0.000 052	0.000 014
2		1.0	0.061 5	0.031 4	0.008 3
3			1.0	0.510	0.134
4				1.0	0.263
All Tank Cars					
Total gallons (millions)	217,583.	364.001	28.556 5	14.474 902	3.245 3
Average gallons per yr (millions)	14,505.533	24.267	1.904	0.965	0.216
Gallons per tank car	19,259.	20,638.	9974.		
Rates 1	1.0	0.001 673	0.000 131	0.000 067	0.000 015
2		1.0	0.078 5	0.039 8	0.008 9
3			1.0	0.507	0.114
4				1.0	0.224

Tank Cars 1980 through 1986					
Nonpressurized Tank Cars					
Total gallons (millions)	90,813.	89.281	6.087 8	3.074 72	1.032 724
Average gallons per yr (millions)	12,973.286	12.754	0.870	0.439	0.148
Gallons per tank car	19,024	19,692.	7,014.	11,346.	10,647.
Rates 1	1.0	0.000 983	0.000 067	0.000 034	0.000 011
2		1.0	0.068 2	0.034 4	0.011 6
3			1.0	0.505	0.170
4				1.0	0.336
Pressurized Tank Cars					
Total gallons (millions)	69,128.	27.478	1.034 2	0.386 8	0.199
Average gallons per yr (millions)	9,875.429	3.925	0.148	0.055	0.028
Gallons per tank car	27,329.	28,225.	10,139.	13,338.	11,056.
Rates 1	1.0	0.000 397	0.000 015	0.000 006	0.000 003
2		1.0	0.037 6	0.014 1	0.007 2
3			1.0	0.374	0.192
4				1.0	0.514
All Tank Cars					
Total gallons (millions)	159,941.	116.759	7.122	3.461 52	1.231 724
Average gallons per yr (millions)	22,848.714	16.680	1.017	0.495	0.176
Gallons per tank car	21,900.	21,200.	7342.		
Rates 1	1.0	0.000 730	0.000 045	0.000 022	0.000 008
2		1.0	0.061 0	0.029 6	0.010 5
3			1.0	0.486	0.173
4				1.0	0.356

Source: Davis and Fowler 1990.

Notes: Rate 1 = gallons carried in or lost from affected tank cars per gallon shipped. Rate 2 = gallons carried in or lost from affected tank cars per gallon in accident damaged tank cars. Rate 3 = gallons carried in or lost from affected tank cars per gallon lost from damaged tank cars with lost lading. Rate 4 = gallons carried in or lost from affected tank cars per gallon lost from punctured tank cars.

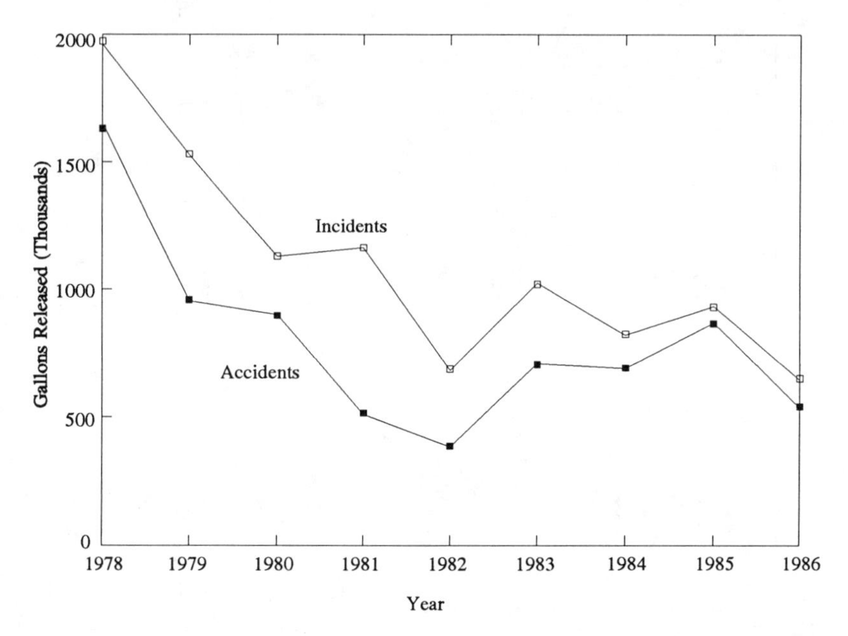

FIGURE 3-21. Gallons released in railroad incidents and corresponding accidents, 1978 through 1986. Source Harvey et al. 1987.

mined for container type directly from the data categorizations. It would be most useful to provide a range of release sizes, and to consider the effects of materials with different volatility and other characteristics potentially able to greatly affect the release amount.

One alternative is to use values recommended by the FEMA/DOT/EPA *Handbook of Chemical Hazard Analysis Procedures* (FEMA et al. undated). For bulk transport by highway, FEMA et al. suggest that 20% of accidents result in a release. Given a release, the FEMA et al. release amount distribution for risk analysis is as follows:

- 60% of the time a 10% cargo loss occurs through a 1-in. hole up to 1,000 gal.
- 20% of the time a 30% cargo loss occurs through a 2-in. hole up to 3,000 gal.
- 20% of the time a complete, instantaneous cargo loss occurs.

If only two release ranges are desired, FEMA et al. recommend using these values for analysis: (1) 80% of the time a 3,000-gal release occurs, and (2) 20% of the time a 10,000-gal release occurs.

For bulk transport by rail, 15% of accidents are said to result in a release. Given a release, the FEMA et al. release amount distribution is as follows:

- 50% of the time a 10% cargo loss occurs through a 2-in. hole up to 3,000 gal.

- 20% of the time a 30% cargo loss occurs through a 2-in. hole up to 10,000 gal.
- 30% of the time a complete, instantaneous loss occurs.

These values are based on a range of container designs and do not distinguish between a cargo of liquid petroleum product or a liquefied gas. Thus, they are useful only for very simplified analyses.

References

Note: The reports of U.S. government agencies, their laboratories, and contractors cited here are available from the National Technical Information Service, Springfield, Virginia 22161, USA.

Abkowitz, M., and G. F. List. January 1986. *Hazardous Materials Transportation: Commodity Flow and Incident/Accident Information Systems.* U.S. Congress. Office of Technology Assessment.

Abkowitz, M., A. Eiger, and S. Srinivasan. 1984. *Assessing the Releases and Costs Associated with Truck Transport of Hazardous Wastes.* PB84-224468. U.S. Environmental Protection Agency.

Bendixen, L. M., and C. P. L. Barkan. 1992. Development of an interindustry risk model for hazardous material transportation by rail. Paper read at the International Consensus Conference: Risks of Transporting Dangerous Goods, April 6–8, University of Waterloo, Toronto, Canada.

BOE (Bureau of Explosives). July 1992a. *Annual Report of Hazardous Materials Transported by Rail, Year 1991.* BOE 91-1. Washington, D.C.: Association of American Railroads.

BOE (Bureau of Explosives). June 1992b. *Report of Railroad Tank Car Leaks of Hazardous Materials by Commodity by Source of Leak for the Year 1991.* BOE 91-2. Washington, D.C.: Association of American Railroads.

Chira-Chavala, T. 1991. Data from TRB-proposed national monitoring system and procedures for analysis of truck accident rates. In *Transportation Research Record 1322*, pp. 44–49. Washington, D.C.: Transportation Research Board.

Clarke, R. K., et al. July 1976. *Severities of Transportation Accidents.* SLA-74-0001. Sandia National Laboratories.

Davis, C. S., and G. F. Fowler. February 1990. *Railroad Tank Car Safety Assessment.* RA-12-4-58 (AAR R-751) Chicago: Association of American Railroads.

Dennis, A. W., et al. May 1978. *Severities of Transportation Accidents Involving Large Packages.* Report SAND77-0001. Sandia National Laboratories.

FEMA (Federal Emergency Management Agency), U.S. Department of Transportation and U.S. Environmental Protection Agency. Undated. *Handbook of Chemical Hazard Analysis Procedures.*

Fischer, L. E., et al. February 1987. *Shipping Container Response to Severe Highway and Railway Accident Conditions.* NUREG/CR-4829. Lawrence Livermore National Laboratory.

FRA (Federal Railroad Administration). June 1987. *Accident/Incident Bulletin No. 155: Calendar Year 1986.* Washington, D.C.: Federal Railroad Administration, Office of Safety.

FRA (Federal Railroad Administration). July 1988. *Accident/Incident Bulletin No. 156: Calendar Year 1987.* Washington, D.C.: Federal Railroad Administration, Office of Safety.

FRA (Federal Railroad Administration). June 1989. *Accident/Incident Bulletin No. 157: Calendar Year 1988.* Washington, D.C.: Federal Railroad Administration, Office of Safety.

FRA (Federal Railroad Administration). June 1990. *Accident/Incident Bulletin No. 158: Calendar Year 1989.* Washington, D.C.: Federal Railroad Administration, Office of Safety.

FRA (Federal Railroad Administration). July 1991. *Accident/Incident Bulletin No. 159: Calendar Year 1990.* Washington, D.C.: Federal Railroad Administration, Office of Safety.

FRA (Federal Railroad Administration). July 1992. *Accident/Incident Bulletin No. 160: Calendar Year 1991.* Washington, D.C.: Federal Railroad Administration, Office of Safety.

Graf, V. D., and K. Archuleta. 1985. *Truck Accidents by Classification.* FHWA/CA/TE-85. U.S. Department of Transportation.

Grimm, J. 1991. Truck safety data needs for the '90s—national truck monitoring system. Paper read at the 1991 Motor Truck Research Symposium, July 10, 1991, Washington, D.C.

Harvey, A. E., P. C. Conlon, and T. S. Glickman. September 1987. *Statistical Trends in Railroad Hazardous Materials Transportation Safety, 1978 to 1986.* R-640. Chicago: Association of American Railroads.

Harwood, D. W., and E. R. Russell. May 1990. *Present Practices of Highway Transportation of Hazardous Materials.* FHWA-RD-89-013. U.S. Department of Transportation.

Harwood, D. W., J. G. Viner, and E. R. Russell. 1990. Truck rate model for hazardous materials routing. In *Transportation Research Record 1264,* pp. 12-23. Washington, D.C.: Transportation Research Board, National Research Council.

Hobeika, A. G., and S. Kim. 1991. Databases and needs for risk assessment of hazardous material shipments by trucks. In proceedings of *Hazmat Transport '91—A National Conference on the Transportation of Hazardous Materials and Wastes,* pp. 2-3-2-30. Evanston, Illinois: Northwestern University.

Hu, P. S., et al. 1989. *Estimating Commercial Truck VMT of Interstate Motor Carriers: Data Evaluation.* ORNL/TM-11278. Oak Ridge, Tennessee: Oak Ridge National Laboratory.

Jovanis, P. P., H. L. Chang, and I. Zabaneh. 1989. A comparison of accident rates for two truck configurations. In *Transportation Research Record 1249,* pp. 18-29. Washington, D.C.: Transportation Research Board, National Research Council.

Moses, L. N., and I. Savage. 1991. Motor carriers of hazardous materials: Who are they? How safe are they? In proceedings of *Hazmat Transport '91—A National Conference on the Transportation of Hazardous Materials and Wastes,* pp. 7-55-7-78. Evanston, Illinois: Northwestern University.

Nayak, P. R., D. B. Rosenfield, and J. H. Hagopian. November 1983. *Event Probabilities and Impact Zones for Hazardous Materials Accidents on Railroads.* PB85-149854. U.S. Department of Transportation.
Neuman, T. R., C. Zegeer, and K. L. Slack. 1991a. *Design Risk Analysis, Volume 1, Final Report.* FHWA-FLP-91-010. Federal Highway Administration.
Newman, T. R., C. Zegeer, and K. L. Slack. 1991b. *Design Risk Analysis, Volume 2, User's Manual.* FHWA-FLP-91-011. Federal Highway Administration.
OTA (Office of Technology Assessment). 1986. *Transportation of Hazardous Materials.* OTA-SET-304. U.S. Congress, Office of Technology Assessment.
Phillips, E. A. March 1990. *Analysis of Aluminum Cars Damaged in Accidents, 1965 through 1986.* RA-20-1-57 (AAR R-749). Chicago: Association of American Railroads.
Phillips, E. A., and H. Role. January 1989. *Analysis of Tank Cars Damaged in Accidents, 1965 through 1986, Documentation Report.* RA-02-6-55 (AAR R-709). Chicago: Association of American Railroads.
RSPA (Research and Special Programs Administration). November 1988. *Transportation Safety Information Report, 1987 Annual Summary.* DOT-TSC-RSPA-88-3. U.S. Department of Transportation.
Saccomanno, F. F., and A. Y. W. Chan. 1985. Economic evaluation of routing strategies for hazardous road shipments. In *Transportation Research Record 1020,* pp. 12–18. Washington, D.C.: Transportation Research Board, National Research Council.
Smith, R. N., and E. L. Wilmot. November 1982. *Truck Accident and Fatality Rates Calculated from California Highway Accident Statistics for 1980 and 1981.* SAND82-7066. Sandia National Laboratories.
Stewart, A. M., and M. Van Aerde. 1990. An empirical analysis of Canadian gasoline and LPG truck releases. *Journal of Hazardous Materials* 25(2): 205–217.
TRB (Transportation Research Board). 1987a. *Designing Safer Roads: Practices for Resurfacing, Restoration, and Rehabilitation.* Special Report 214.
TRB (Transportation Research Board). 1987b. *Relationship Between Safety and Key Highway Features: A Synthesis of Prior Research.* State of the Art Report 6.
TRB (Transportation Research Board). 1990. *Special Report 228: Data Requirements for Monitoring Truck Safety.* Washington, D.C.: Transportation Research Board, National Research Council.

4

Development of Accident Scenarios

The objective of this chapter is to show how accident scenarios are developed using fault trees and event trees. The chapter provides fault trees and event trees that can be used directly, or with modification, for many transportation QRAs.

4.1 FAULT TREE ANALYSIS

A fault tree is a graphical presentation of the systematic, logical development of the many causes of an undesirable event. Fault trees have been used extensively in transportation risk analysis (Section 2.3). Fault tree analysis begins with an undesirable event called the top event, for example, "release of hazardous material" or "container failure from impact." The next step is to identify the immediate, necessary, and sufficient causes of the top event. Each of the immediate causes is examined in turn to determine its immediate causes. This process continues until the analyst has obtained the desired level of resolution of causes. Each cause will later be assigned probabilities or frequencies from the available data.

If we use chlorine as an example for the hazardous material being transported, the top event might be "release of chlorine from tank car during transport operations." The next step is to ask "What are the direct causes of release?" or "How can this happen?" The analyst could list several possible direct causes (Andrews et al. 1980):

- Chlorine released from tank head during transport.
- Chlorine released from tank shell during transport.
- Chlorine released from manway cover during transport.
- Chlorine released from valves during transport.

The causes for "chlorine released from tank head during transport" might be (Andrews et al. 1980):

- Chlorine released from tank head during normal transport.
- Chlorine released from tank head during transportation accident.

The analyst would continue to develop causes for these two possibilities.

The sequence of causes of the top event is not unique. An alternative to the preceding is first to list the following possible causes of the top event, "release of chlorine from tank car during transport operations":

- Chlorine released from loading or unloading operations.
- Chlorine released from normal transport.
- Chlorine released from storage or other operations that are subject to natural phenomena or other external events.
- Chlorine released from transport accidents.

This list of causes has the potential advantage of separating the phase of transport operations at an early stage. Each of these four causes could be treated as the top event for four separate fault trees, or the development could continue for a single, combined tree. In this book only the releases from normal transport and from transport accidents are addressed, with emphasis on the latter.

To continue to develop causes for "chlorine release from transport accidents," the analyst should be aware of the available data and of course the objectives of the analysis. The next set of causes might be the following:

- Impact forces fail tank car.
- Crush forces fail tank car.
- Puncture forces fail tank car.
- Thermal forces fail tank car.
- Tank car damaged by impact subsequently fails from thermal forces.

This approach is tailored for assignment of probabilities using data given in Section 3.2.

The process for developing the tree has been described previously. The tree is constructed by using the logic symbols shown in Fig. 4-1, and the immediate causes of the top event and the transport release event just described are shown in Figs. 4-2 and 4-3. (The reference to chlorine has been deleted in the interest of greater generality.) The tree will be developed further in Section 4.2. Additional details on fault tree construction are described in several references, for example, Roberts et al. (1981), McCormick (1981), and Henley and Kumamoto (1981).

The completed fault tree displays the combinations of events that are required to result in the top event. The relatively simple fault trees in this

Name	Symbol	Function
Top event Intermediate event	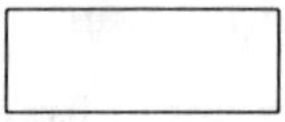	Describes a fault event resulting from the combination of more basic faults
OR gate		Denotes that one or more of the input events must exist for the output event to occur
AND gate		Denotes that all of the input events must exist simultaneously for the output event to occur
Inhibit condition	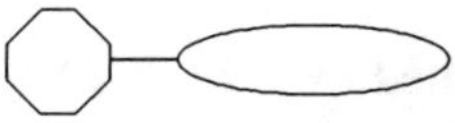	Denotes that the output event occurs when the input event occurs and the inhibit condition (in oval) is satisfied
Basic event		Denotes a basic component failure or condition that is usually assumed to be independent of other basic and undeveloped events
Transfer out	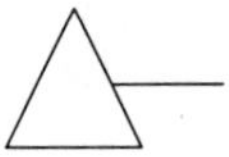	Denotes that this part of the fault tree is a further development or duplication of another part of the fault tree
Transfer in		Denotes that the fault tree is developed further elsewhere on the tree

FIGURE 4-1. Fault tree logic symbols.

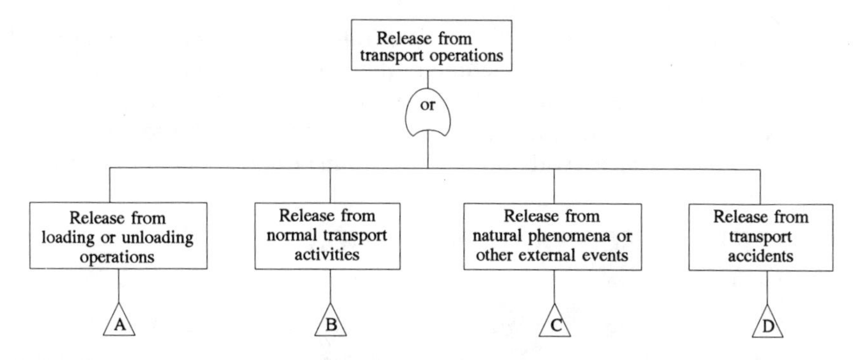

FIGURE 4-2. Fault tree for releases from transportation operations.

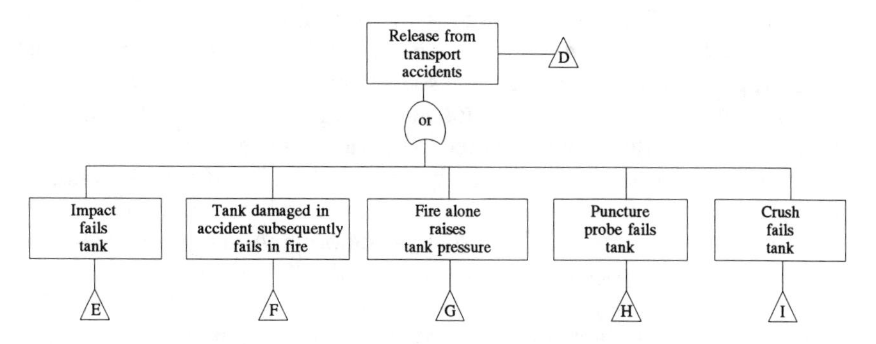

FIGURE 4-3. Fault tree for releases from transportation accidents.

book are developed such that the accident scenarios, that is, the combinations of events, that are sufficient to cause the top event can be identified directly from the tree. In most applications, however, even an experienced analyst cannot identify directly all of the combinations that can occur on the fault tree. A complex fault tree is solved by using computer programs to determine the combinations of events and/or conditions that are necessary and sufficient to cause the specific top event. The qualitative solution (listing of events that constitute accident scenarios) can provide powerful insight into a complex system. When probabilities and frequencies are assigned to the basic events, the resulting quantitative solution provides the frequency or probability for each set of events and for the occurrence of the top event.

Simple fault trees can be quantified by hand calculations by starting at the bottom and working upward. The input probability values of an AND gate are multiplied to obtain the output probability, and the input probability values of an OR gate are summed to obtain the output probability. The inhibit condition is zero if the inhibit condition is not satisfied, or the product of the input probability and the inhibit probability if the condition is satisfied. If all input values are probabilities, the output is a probability. A probability times a frequency is a frequency, but a probability cannot be added to a frequency. Two frequencies with the same units can be added but not multiplied.

4.2 GENERALIZED TRANSPORTATION FAULT TREE

In this section, a fault tree will be constructed for the chlorine tank car shown in Fig. 4-4 (GATX 1985). This fault tree is a significant modification of the one presented by Andrews et al. (1980). It usually is possible to modify and/or reduce the fault tree constructed in this section for application in other situations. The chlorine tank car will be treated as a "typical" rail tank car. Clearly, for extension to other tank cars, the presence of bottom outlet valves, rupture disks, steam lines, and so forth must be accounted for. The car shown in Fig. 4-4 is loaded and unloaded through valves located in the tank dome. Two 1-in. valves are connected to the vapor space, and each of the two additional 1-in. valves is connected to the bottom of the liquid space by a 1.25-in. pipe. An excess flow valve located just below each liquid valve will close if the rate of flow exceeds 7000 lb/hr of liquid chlorine. A safety relief valve will start to open at 375 psig and has a rated flow of 1.383 m^3/s (2930 scfm of air) at full open condition, which occurs at 412 psig (The Chlorine Institute 1979).

The causes of "impact fails tank" in Fig. 4-3 are given in the second line of Fig. 4-5. To determine the cause of "impact fails tank liquid valves and excess flow valve does not close," the first AND gate is encountered. Not only must the tank liquid valves fail from the impact, but the excess flow valves also

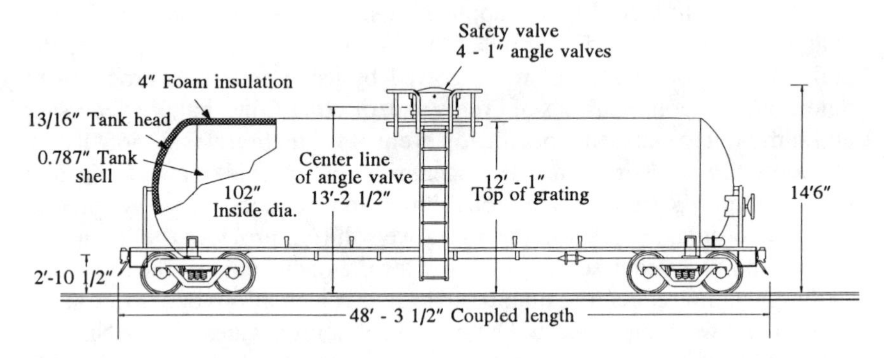

FIGURE 4-4. A 90-ton capacity, DOT 105A500W railcar for chlorine service (post-1982). Source: GATX *Tank Car Manual* 1985.

must fail to close. (The small amount of liquid that would be released before the liquid valves close is neglected in this case.) Under "impact fails tank liquid valves," the first "inhibit condition" is encountered. Even if "impact occurs on tank dome" takes place, failure only results if the impact is of sufficient magnitude to fail the liquid valves.

The remainder of Fig. 4-5 is developed in a similar manner. Note that the possibility that the shell, the head, or the manway is defective has been included. Defects could have been introduced by initial faulty welding, the use of defective materials, or the gradual degradation of wall thickness or welds. A boiling-liquid, expanding-vapor explosion can be initiated by impact (see Section 6.1.1). This possibility could be incorporated into the "impact sufficient to fail shell" inhibit condition in the tree as constructed. An explicit cause could also be added under the "impact fails tank" event.

Many of the basic events of Fig. 4-5 incorporate the accident frequency and the conditional probability of an accident force type. For example, to evaluate "impact occurs on tank dome," three parameters are multiplied together:

- The frequency of an accident.
- The probability of an impact force occurring, given an accident.
- The probability that the tank dome experiences the impact force.

However, to explicitly and continuously illustrate on the tree the accident frequency and force type probability, that is, to show them as basic events, would unnecessarily clutter the figure.

The way that Figs. 4-2 through 4-5 are constructed provides the analyst with the ability to easily see the various accident scenarios that lead to the top event. Choosing the middle of Fig. 4-5 as an example, the following things happen:

- An accident occurs.
- An impact force occurs.
- The impact force is exerted on a shell with a defect.
- The force is sufficient to fail the shell.

This sequence follows the first few blocks on the top line of Fig. 2-3 and the first few terms of Eq. 2-2.

Figures 4-6 and 4-7 are constructed in much the same way as Fig. 4-5. A release caused only by thermal forces, the top event in Fig. 4-8, can be caused by a release through the functioning relief valve or because the tank body fails. A vapor release through the relief valve will occur if the tank is upright (or nearly so) and the fire is of sufficient duration to raise the pressure to the relief valve setpoint. Test data (Townsend et al. 1974, as cited in Johnson 1984) show that if the thermal driving force continues until the tank car is about one-half full, the tank car shell will fail. When the tank car is nearly full, convective heat transfer by the liquid contents cools the inner tank car

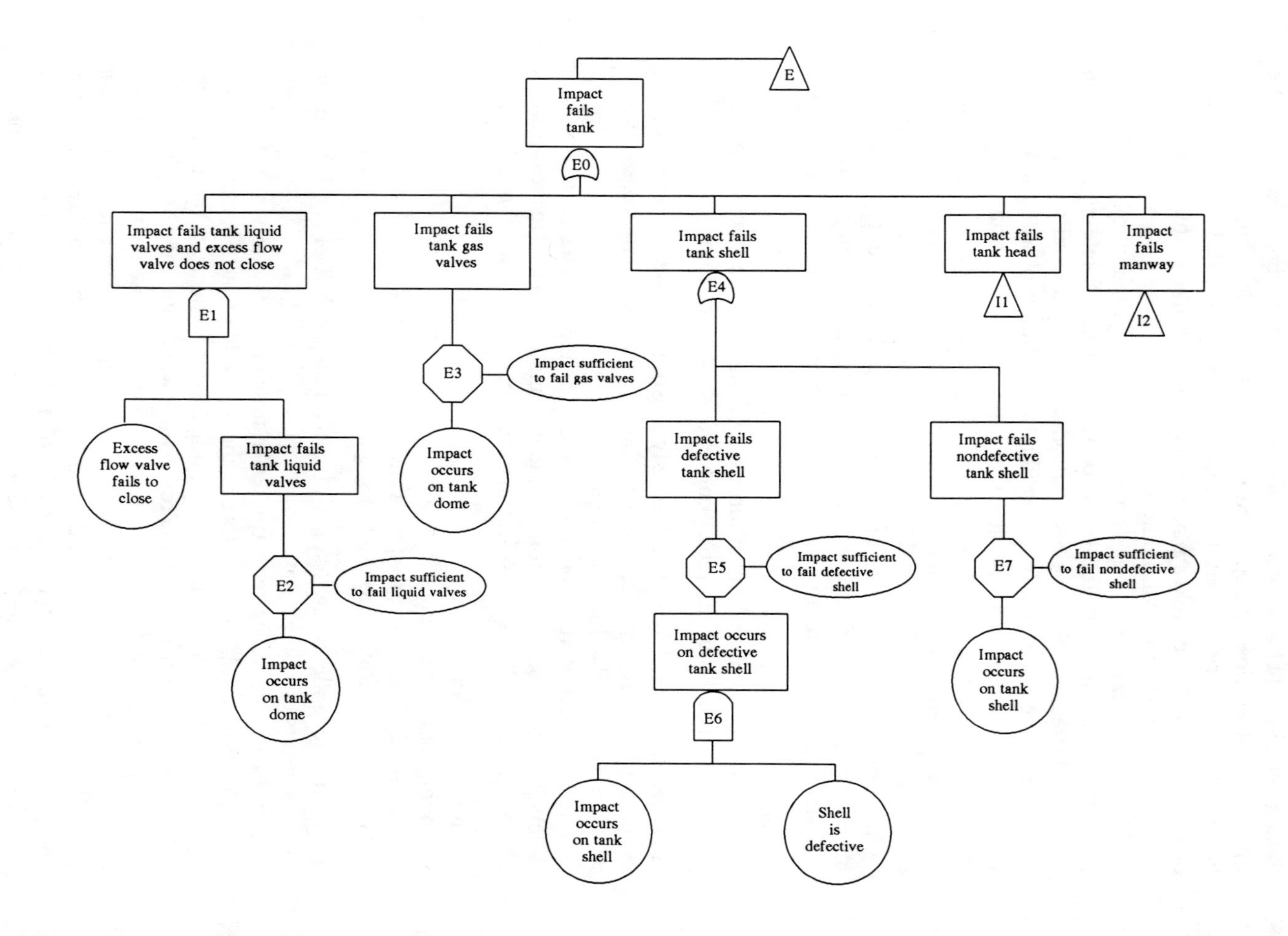
E
Impact fails tank
E0
Impact fails tank liquid valves and excess flow valve does not close
Impact fails tank gas valves
Impact fails tank shell
Impact fails tank head
Impact fails manway
E1
E3
E4
I1
I2
Impact sufficient to fail gas valves
Excess flow valve fails to close
Impact fails tank liquid valves
Impact occurs on tank dome
Impact fails defective tank shell
Impact fails nondefective tank shell
E2
Impact sufficient to fail liquid valves
Impact occurs on tank dome
E5
Impact sufficient to fail defective shell
E7
Impact sufficient to fail nondefective shell
Impact occurs on defective tank shell
Impact occurs on tank shell
E6
Impact occurs on tank shell
Shell is defective

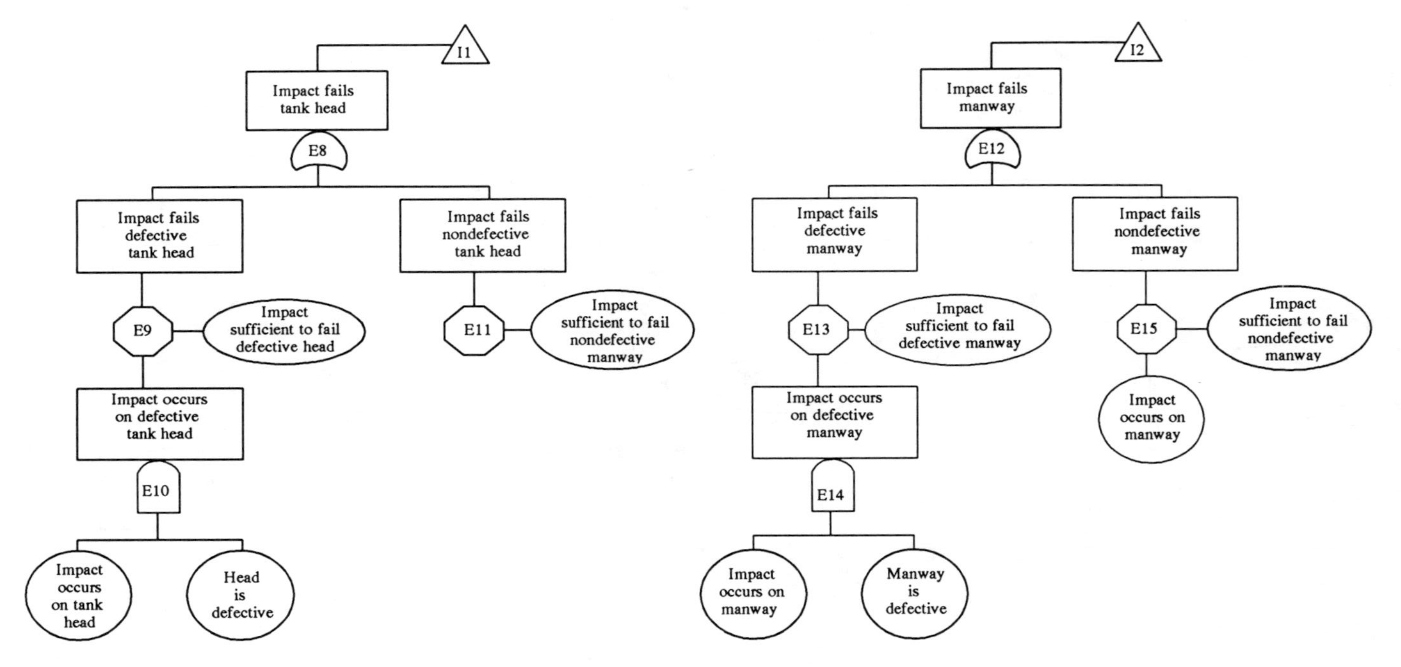

FIGURE 4-5. Fault tree for impact failures.

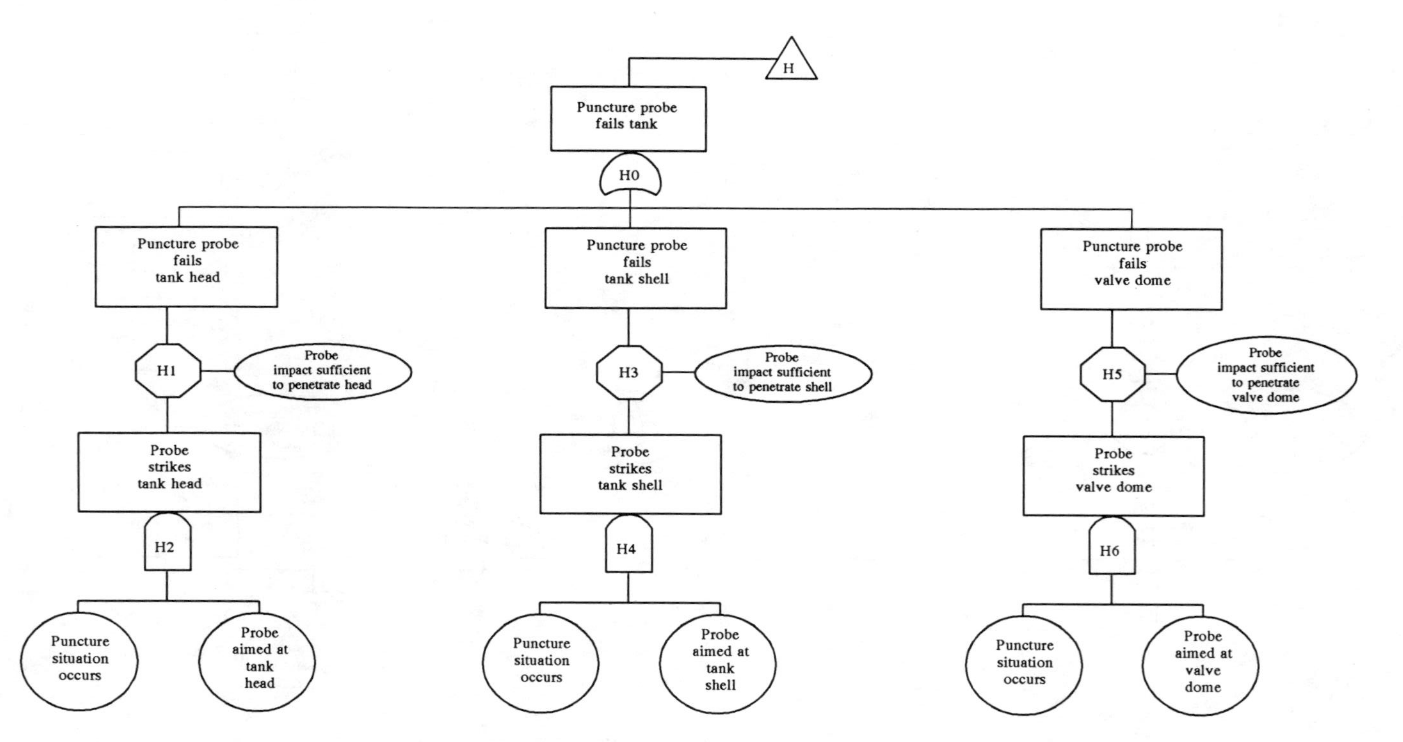

FIGURE 4-6. Fault tree for puncture failures.

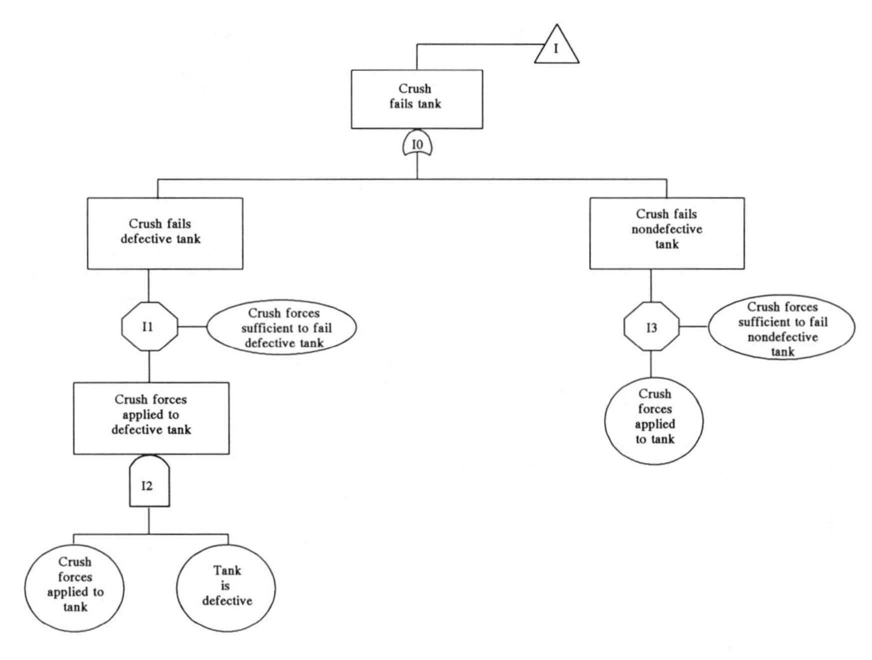

FIGURE 4-7. Fault tree for crush failures.

wall. At about the 50% full condition, the reduced heat transfer capability of the vapor phase results in vapor space wall temperature increases to such a level that the wall strength is no longer capable of maintaining the tank's integrity. Fire durations capable of causing tank failure from this mechanism are addressed under the "fire causes ejection of greater than 50% of tank inventory" event. The presence of insulation can substantially delay tank failure times (Townsend et al. 1974, cited as in Johnson 1984). The tank car contents still are released, but the release duration is lengthened. A boiling-liquid, expanding-vapor explosion could result (see Section 6.1.1).

Liquid release from the relief valve does not remove as much energy from the tank car as does vapor release; therefore, the tank car pressure may continue to rise if the relief valve setpoint is reached and the car is upside down (or nearly so). The potential exists to fail the tank car by overpressure even though the relief valve has opened. If the "fire duration sufficient to raise tank pressure to relief valve set pressure but not to overpressurize tank with only liquid discharge" inhibit condition is satisfied, then the "liquid release occurs" event takes place. If the "fire duration sufficient to overpressurize tank with only liquid discharge" inhibit condition is met, the "fire pressurizes tank above relief valve set pressure" event takes place instead.

Figure 4-9 is constructed in the same way as Fig. 4-8 with the added

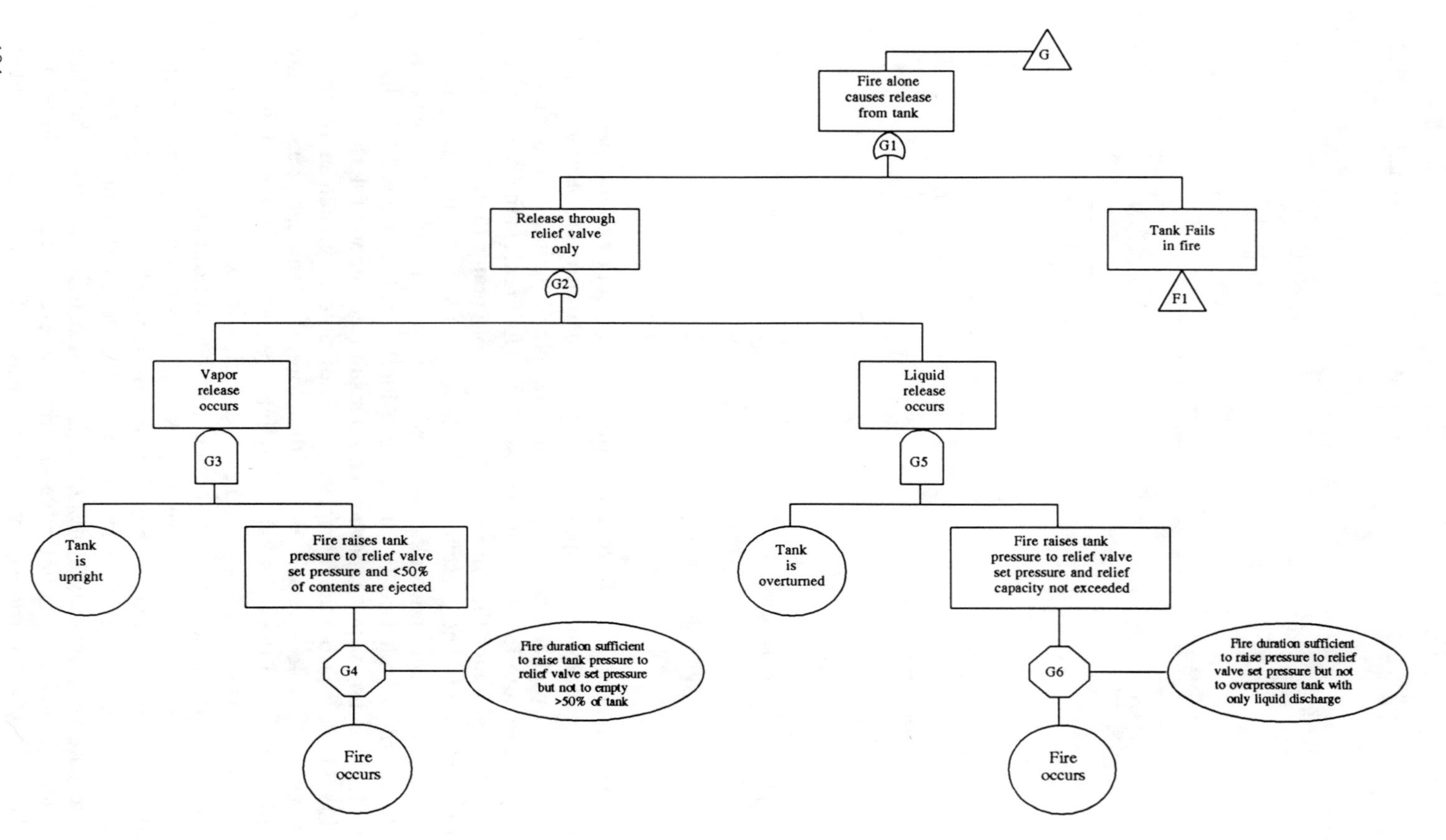

G
Fire alone causes release from tank
G1
Release through relief valve only
Tank Fails in fire
F1
G2
Vapor release occurs
Liquid release occurs
G3
G5
Tank is upright
Fire raises tank pressure to relief valve set pressure and <50% of contents are ejected
Tank is overturned
Fire raises tank pressure to relief valve set pressure and relief capacity not exceeded
G4
Fire duration sufficient to raise tank pressure to relief valve set pressure but not to empty >50% of tank
Fire occurs
G6
Fire duration sufficient to raise pressure to relief valve set pressure but not to overpressure tank with only liquid discharge
Fire occurs

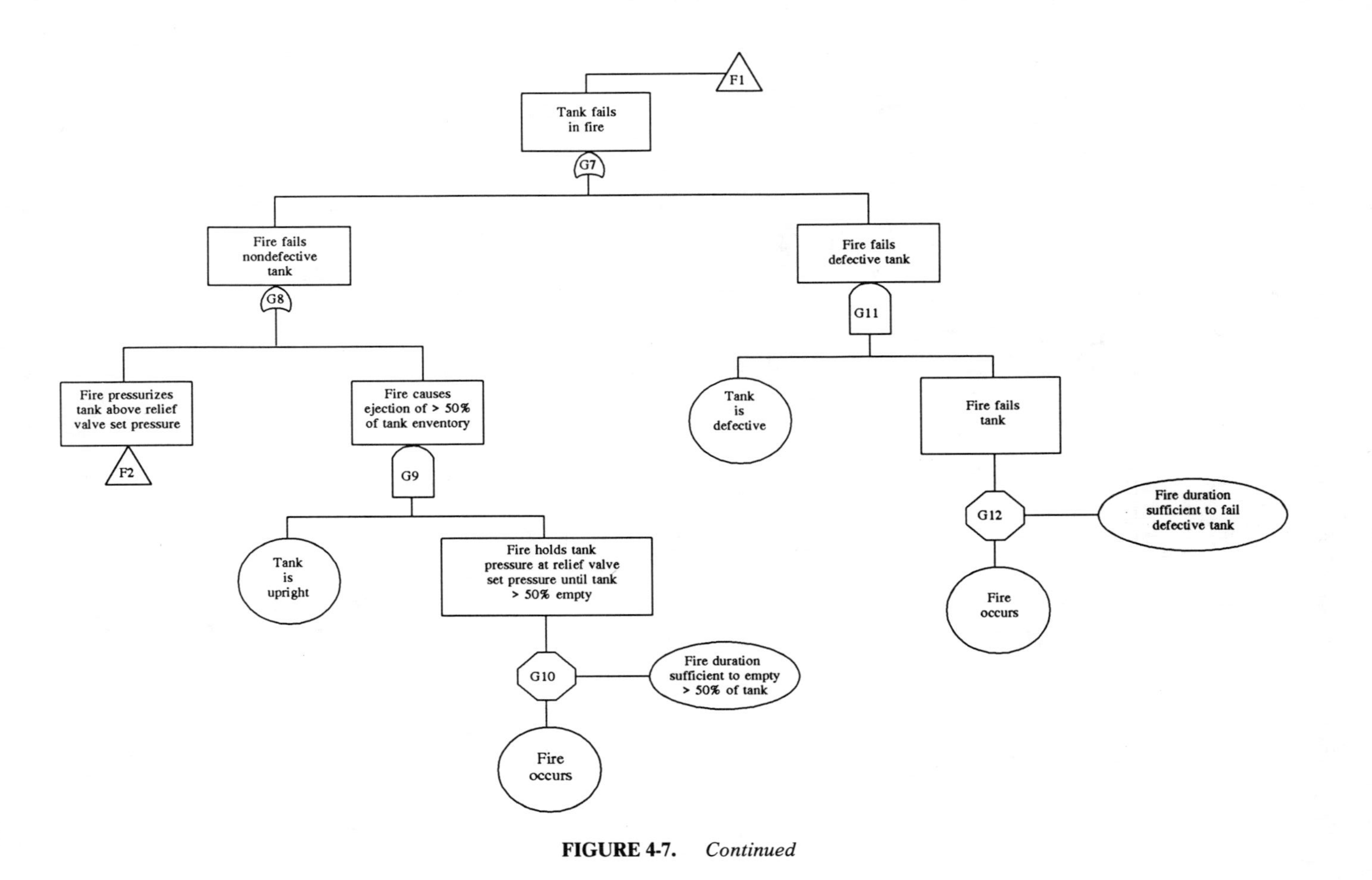

FIGURE 4-7. *Continued*

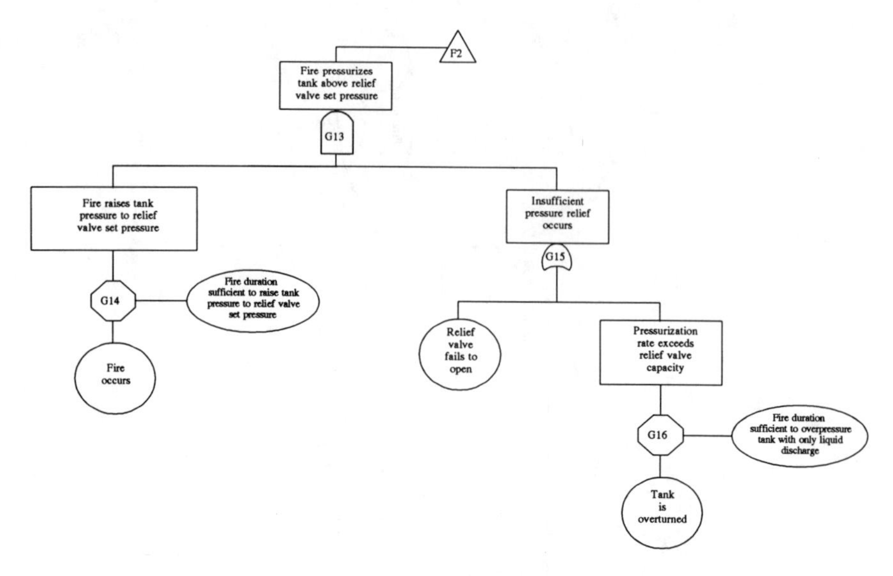

FIGURE 4-8. Fault tree for failures caused only by fire.

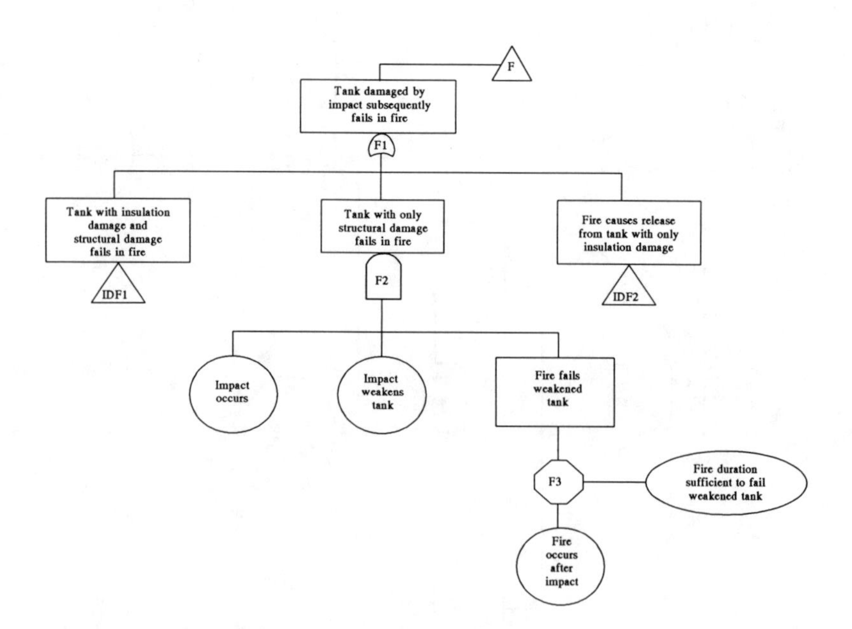

FIGURE 4-9. Fault tree for mechanical damage followed by failure from fire.

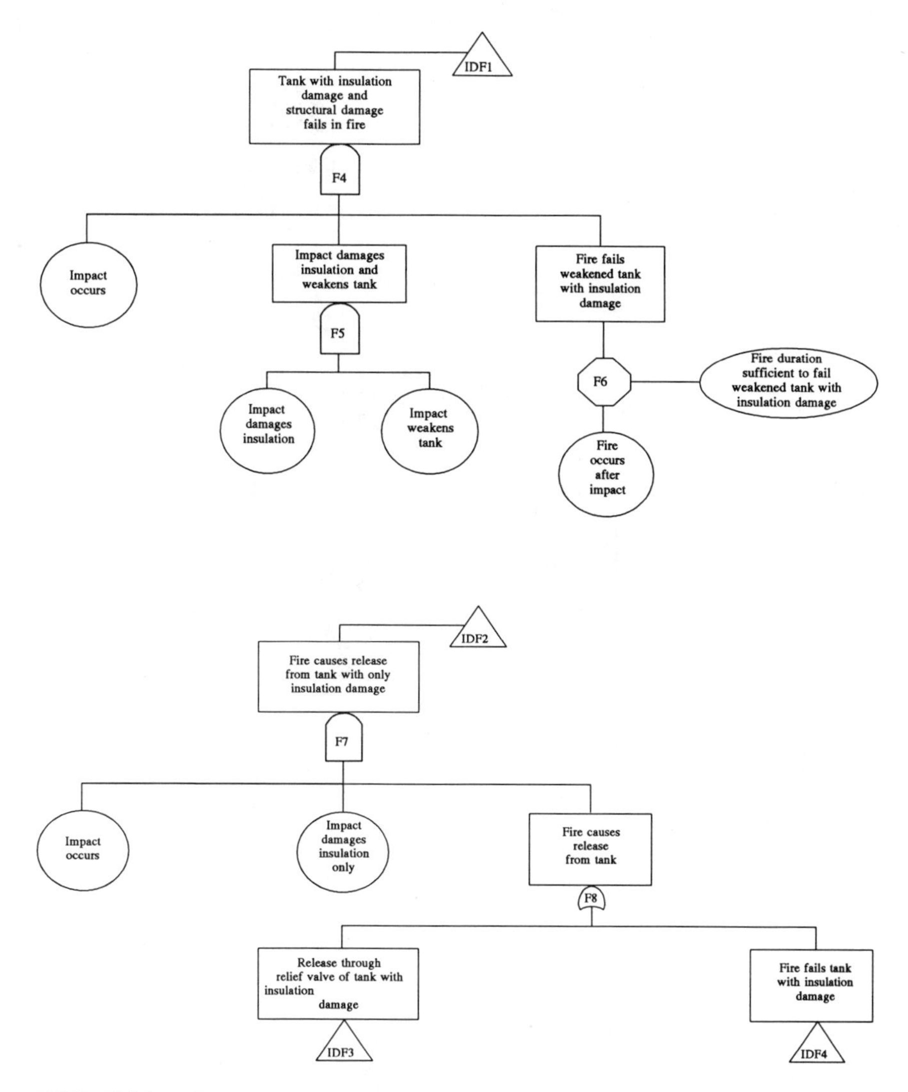

FIGURE 4-9. *Continued*

complication that the tank car has suffered impact damage to the insulation, and/or the impact has weakened, but not failed, the tank car body.

Figures 4-5 through 4-9 are general and apply directly to many tank cars and tank trucks. In some cases, the gas and liquid valve portions will need to be modified to incorporate the feature of a bottom outlet valve or other features. Ton containers can be addressed by deleting references to manway

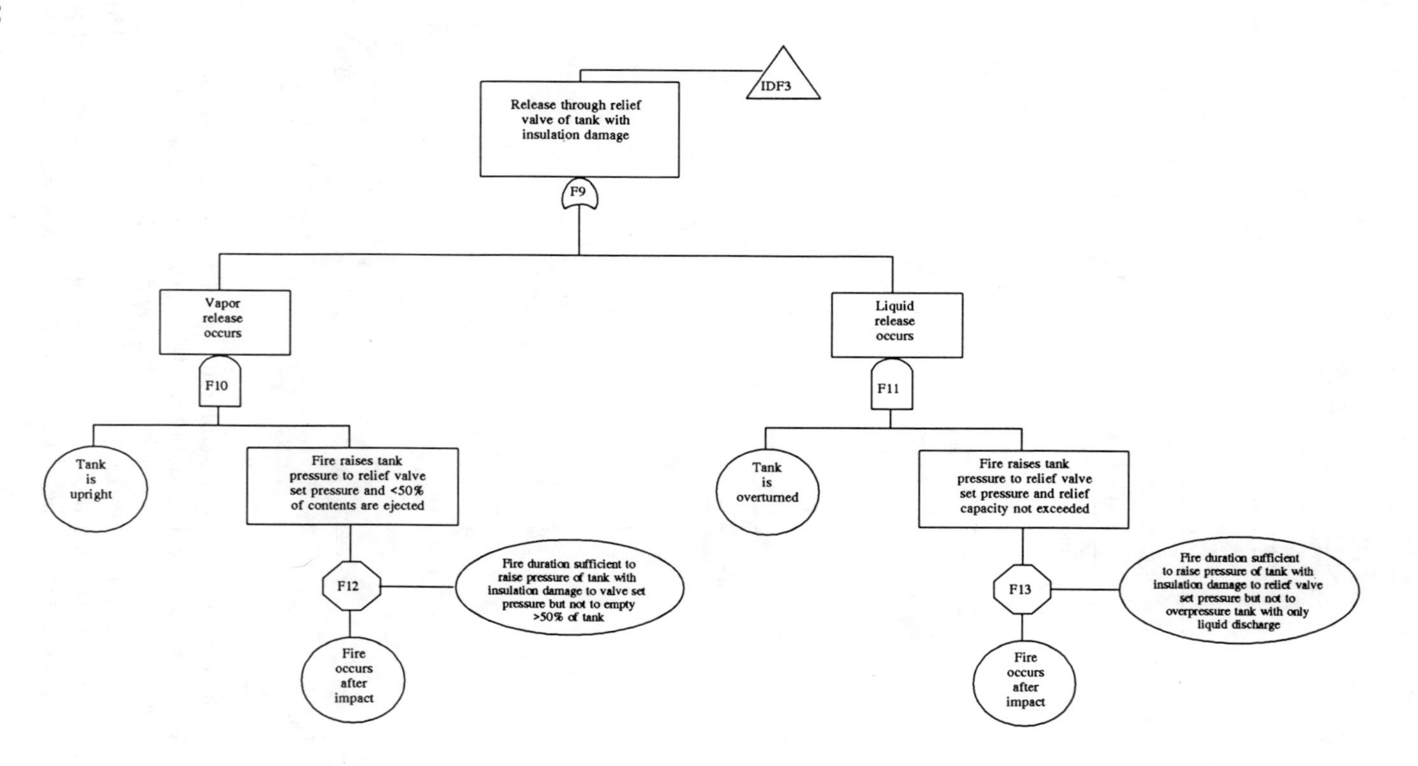

IDF3
Release through relief valve of tank with insulation damage
F9
Vapor release occurs
F10
Liquid release occurs
F11
Tank is upright
Fire raises tank pressure to relief valve set pressure and <50% of contents are ejected
F12
Fire duration sufficient to raise pressure of tank with insulation damage to valve set pressure but not to empty >50% of tank
Fire occurs after impact
Tank is overturned
Fire raises tank pressure to relief valve set pressure and relief capacity not exceeded
F13
Fire duration sufficient to raise pressure of tank with insulation damage to relief valve set pressure but not to overpressure tank with only liquid discharge
Fire occurs after impact

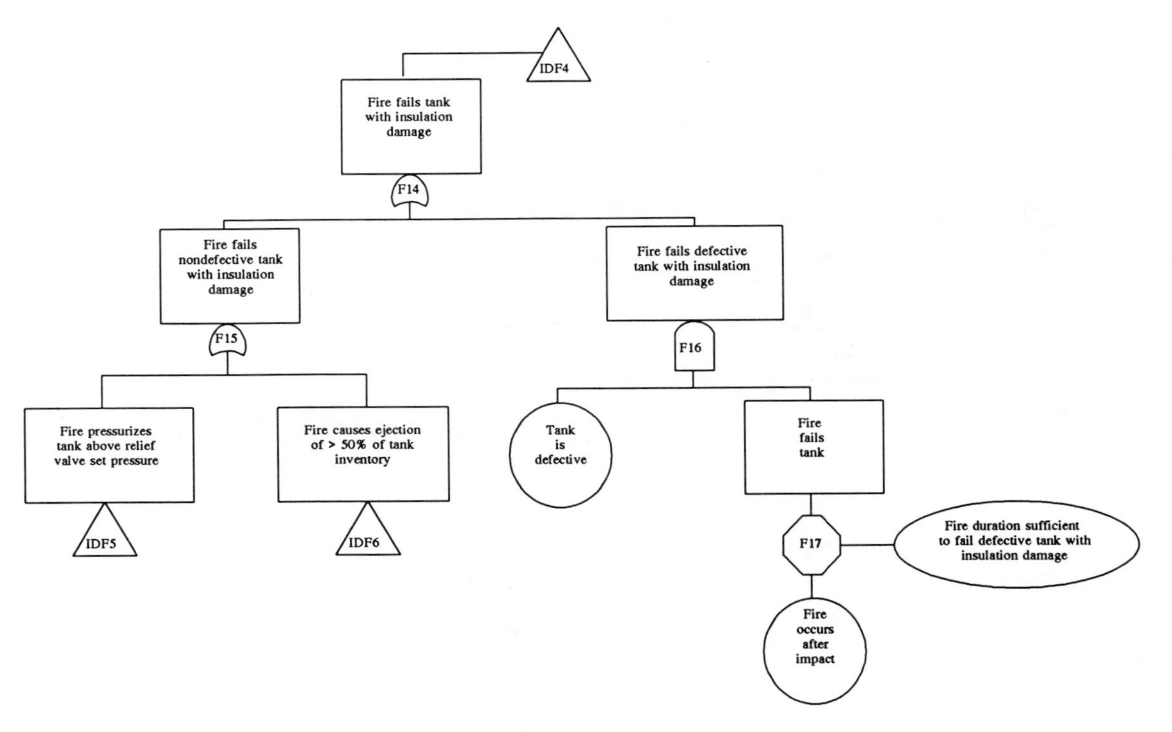

FIGURE 4-9. *Continued*

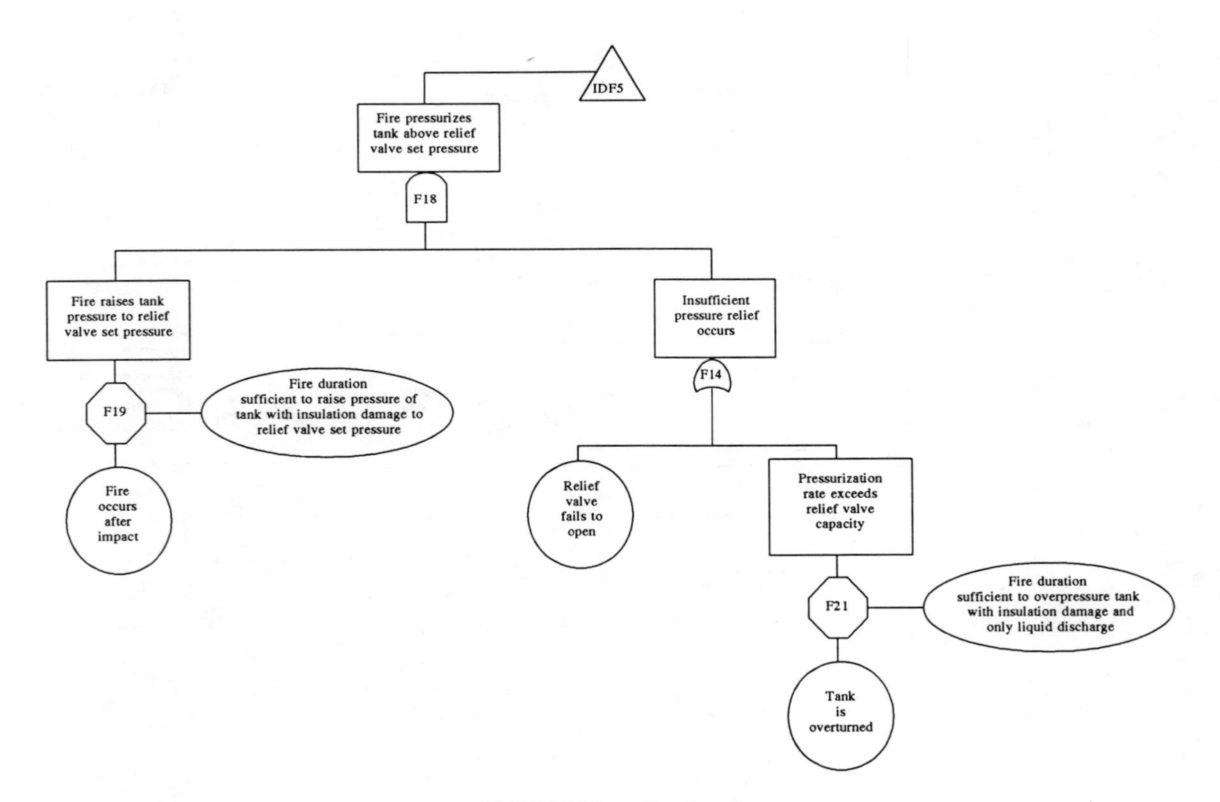

FIGURE 4-9. *Continued*

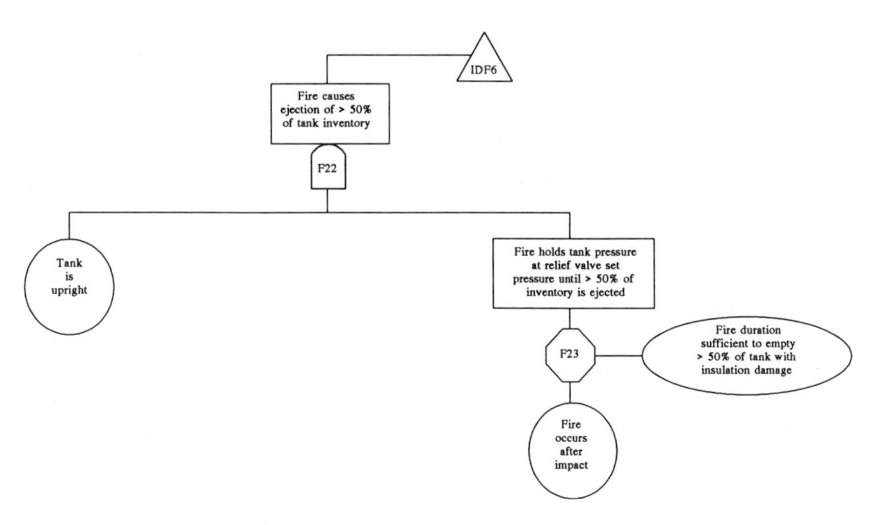

FIGURE 4-9. *Continued*

and insulation and considering the tank dome to be the bonnet, and by changing orientation requirements for flow from the gas/liquid spaces.

4.3 EVENT TREE ANALYSIS

An event tree is a graphical model for identifying and evaluating potential outcomes from a specific initiating event. The event tree depicts the chronological sequence of events, that is, accident scenarios, that could result from the initiating event.

The first step is to identify the initiating event, for example, "accident occurs producing crush force." The analyst asks what protective system action, operator action, normal system function, and so forth is expected to occur next. Each event following the initial event is conditional on the preceding event. The outcomes of the events usually are binary, that is, the success or failure of an operator action or, alternatively, "yes, the safety system functioned," or "no, it did not." An example of a nonbinary or multiple outcome is positioning of a control valve at 100%, 50%, 30%, or 0% of fully open.

For the initiating event just defined, the analyst may be tempted to assign as the next event "tank fails from crush force"; however, in the development of the fault tree the possibility of a defective tank car wall was considered. To include that possibility, the next event must be "tank is defective." Then the "tank fails from crush force" event can be added.

To construct the tree, a horizontal line is drawn on the left-hand side of a page, and the accident initiator is identified directly above at the top of the sheet (Fig. 4-10). The next event is listed at the top and to the right of the initiating event, and the binary outcome of the event is indicated by a branch point that splits the initiating event into two states indicated by two horizontal lines. As shown in Fig. 4-10, the "tank is defective" condition is the upper branch, and the "tank is not defective" condition is the lower branch. For each of these two branches, the "tank fails from crush force" event applies, and the two branches each split to form a total of four branches. At this point four accident scenarios have been defined, one for each of the four branches. Two branches are for two tank failure scenarios, and two are for scenarios in which tank integrity is maintained.

Data from Chapter 3 indicate that fire can occur after events involving crush forces. If the tank car has failed from crush forces, the fire still can affect the release characteristics. If crush forces have not failed the tank car, this event tree, by design, will not address the possibility that the fire is of sufficient length to fail the tank car. Thus the event "fire occurs" adds new information only to the two failure scenarios, and the fact that it is not applicable to the two nonfailure scenarios is shown by the lack of a branch point in these scenarios on Fig. 4-10.

Event trees sometimes are classified as having preincident or postincident application (CCPS 1989). The preincident tree addresses the events up to

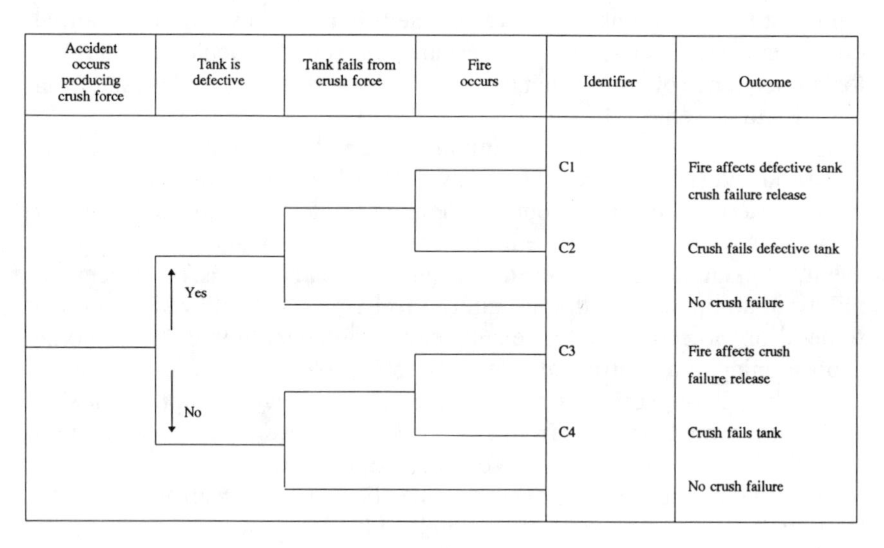

FIGURE 4-10. Event tree for crush force.

hazardous material release, and the postincident tree addresses factors that affect the range of consequences. By this definition, the "fire occurs" column is not needed in Fig. 4-10; however, experience has shown that fire following mechanical failure often is overlooked during consequence calculations unless it is part of the event tree. Furthermore, because data with which to estimate the fire probability are available, the event is useful in the preincident tree; so the quantification of the scenario frequency is completed by using all of the data from Chapter 3 in one pass.

A column usually is added to the event tree after the last event to insert a unique accident sequence identifier for later use. Usually only failure consequences are labeled. A column for labeling the outcomes of the sequences also is very useful, as shown in Fig. 4-10.

By convention, the event tree usually is arranged so that the outcomes trend from less severe at the top to more severe at the bottom of the outcome column. This convention is convenient for developing trees that address the ability of operator actions and active safety systems to prevent the release of hazardous material. Success or "yes" in the upward branching direction will put the most benign outcomes at the top for such events. For passive systems such as tank cars, it frequently is awkward to word the event descriptors as success events to achieve the same ordering of outcomes. For example, in Fig. 4-10, the failure event, "tank fails from crush force," labeled as a success event would become "tank does not fail from crush force" and the yes/no is read as "yes, the tank does not fail from crush force" and "no, the tank does not fail from crush force." Convention could be followed by describing the event as "tank survives the crush force," but the author believes that less confusion occurs by using failure event descriptions across the top of the event tree. Because the outcomes all are labeled, there is little potential for confusion just because the most benign outcome is at the bottom. Additional details and examples of event tree construction are provided in several references, for example, Henley and Kumamoto (1981) and CCPS (1989).

4.4 GENERALIZED TRANSPORTATION EVENT TREES

The analogous event trees for puncture force, fire force (only), and impact force initiators are presented in Figs. 4-11 through 4-13. The way that fault trees can be used to provide quantitative input into an event tree is illustrated in Figs. 4-11 through 4-13. (The defective tank possibility is included in the fault tree.) In Fig. 4-11, for example, the fault tree of Fig. 4-6, designated by

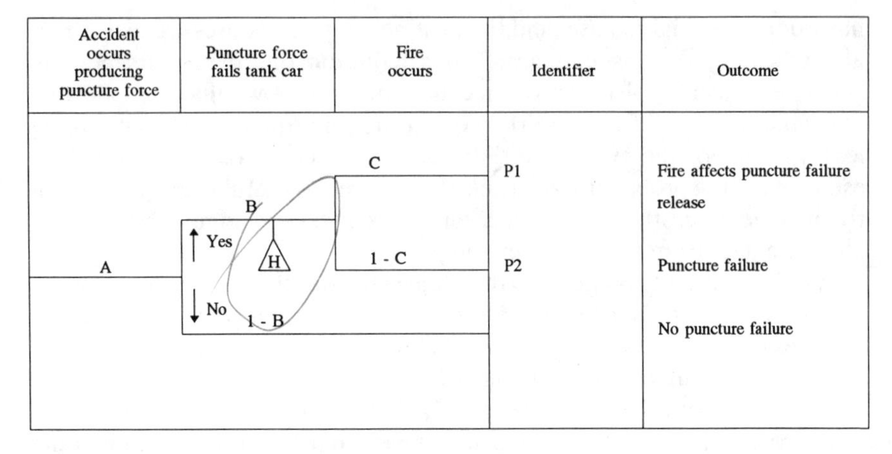

FIGURE 4-11. Event tree for puncture force.

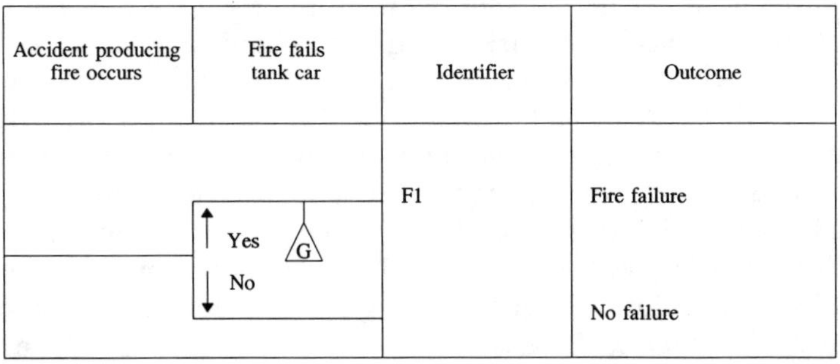

Accident producing fire occurs	Fire fails tank car	Identifier	Outcome
	Yes (G)	F1	Fire failure
	No		No failure

FIGURE 4-12. Event tree for fire force only.

the H transfer, is evaluated quantitatively to determine the probability, indicated by the quantity B, that puncture forces fail the tank car. The probability that puncture forces do not fail the car is 1 minus that probability, or $1 - B$. The frequency or probability of sequence $P2$ on Fig. 4-11 is A times B times the quantity $1 - C$. To use the H transfer from Fig. 4-6 in this way, the fault tree basic event, "puncture situation occurs," must be quantified so that the quantification of "accident occurs producing puncture forces" in the event tree is consistent. The same is true for the other fault tree/event tree combinations, which is a major reason why the fault tree basic events are defined as they are in Section 4.2.

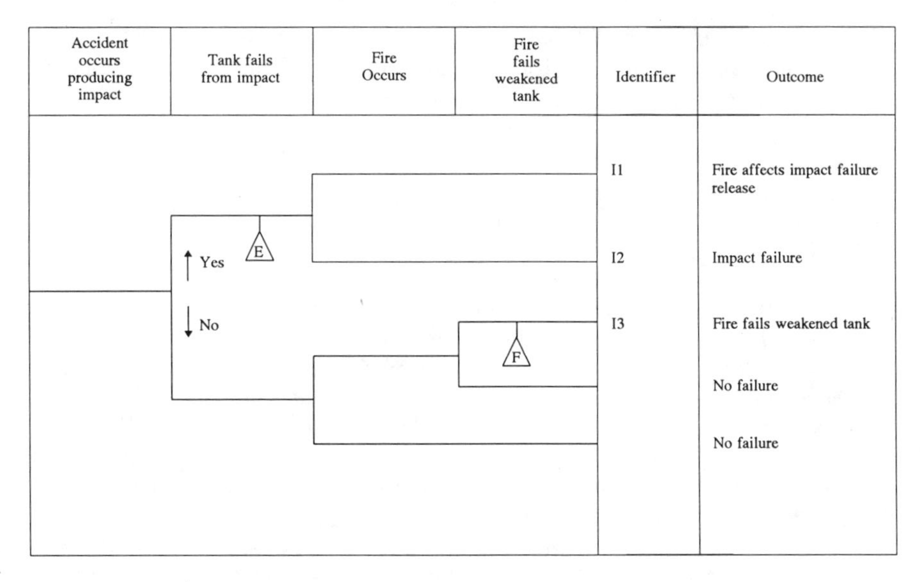

FIGURE 4-13. Event tree for impact force.

4.5 SUMMARY

As with all aspects of QRA, fault trees and event trees can be constructed to many levels of detail. For example, a valve can be described as if it is a single piece of equipment, or it can be described as a system consisting of a valve body and other moving and stationary parts.

Fault trees and event trees are widely used in QRAs for transportation and for stationary chemical and nuclear facilities. They are powerful tools when used qualitatively to give insight into how a system will function in an abnormal situation. Risk reduction measures can be effectively deduced with this information alone. When quantified, either a fault tree or an event tree can provide the overall frequency of an accident scenario of interest. The fault tree can be used to provide quantitative input to an event tree, or it can be used alone.

The causes of transport accidents involve a variety of equipment failures, operator failures, and conditions external to the transport vehicle, such as weather, road conditions, and other drivers. Modeling of the failures, including interactions between the equipment, the driver, and the external environment, is not well enough developed to predict accident rates. However,

Joshua and Garber (1992) have used fault tree analysis to model these interactions on a system level to help explain the interactions and to help suggest effective risk reduction measures.

References

Note: The reports of U.S. government agencies, their laboratories, and contractors cited here are available from the National Technical Information Service, Springfield, Virginia 22161, USA.

Andrews, W. B., et al. March 1980. *An Assessment of the Risks of Transporting Liquid Chlorine by Rail.* PNL-3376. Pacific Northwest Laboratory.

CCPS (Center for Chemical Process Safety). 1989. *Guidelines for Chemical Process Quantitative Risk Analysis.* New York: American Institute of Chemical Engineers.

GATX (General American Transportation Corporation). Fifth Edition, February 1985. *Tank Car Manual.* Chicago: General American Transportation Corporation.

Henley, E. J., and H. Kumamoto. 1981. *Reliability Engineering and Risk Assessment.* Englewood Cliffs, New Jersey: Prentice-Hall.

Johnson, M. R. June 1984. *Temperatures, Pressures and Liquid Levels of Tank Cars Engulfed in Fires, Volume 1, Results of Parametric Analyses.* DOT/FRA/OR&D-84/08.I (available as PB85-156859). Federal Railroad Administration.

Joshua, S. C., and N. J. Garber. 1992. A causal analysis of large vehicle accidents through fault-tree analysis. *Risk Analysis* 12(2):173–187.

McCormick, N. J. 1981. *Reliability and Risk Analysis.* New York: Academic Press.

Roberts, N. H., et al. 1981. *Fault Tree Handbook.* NUREG-0492. U.S. Nuclear Regulatory Commission.

The Chlorine Institute. March 1979. *Chlorine Tank Car Loading, Unloading, Air Padding, Hydrostatic Testing.* Edition 1. Washington, D.C.: The Chlorine Institute.

Townsend, W., et al. December 1974. *Comparison of Thermally Coated and Uninsulated Rail Tank Cars Filled with LPG Subjected to a Fire Environment.* FRA-OR&D-75-32. Federal Railroad Administration.

5

Engineering Models for Container Failure Analysis

The historical and predictive approaches to transportation risk methodology are described in Section 2.4. The historical approach relies on data presented in Section 3.3, and the predictive approach relies on the analyses described in this chapter.

Potential approaches for determining the magnitude of accident forces required to produce container failure are summarized in this chapter. (Non-accident releases are not pertinent here.) The objective is to give the potential transportation risk project manager a feeling for types of analyses to consider in planning a project. Competent structural and thermal analysts can respond to the project manager's general guidance for the specific commodity/container under study. Results and analytical approaches are described for selected containers, but detailed presentation of mechanical and thermal analysis techniques is beyond the scope of this book.

The purpose of most testing and analysis of transport containers is to show that the container will survive a particular threat specified in federal or state regulations, not to find the minimum threat magnitude that produces failure (failure threshold); so many tests and analyses are of limited value.

It is impossible to pinpoint failure thresholds. Engineering estimates, therefore, usually are deliberately conservative.

5.1 ANALYSES REQUIRING FAILURE MODELS

Data sources for the conditional probability of container failure for the historical approach are described in Section 3.3. If the historical approach is used for determining container failure, given an accident, then no additional engineering analysis is needed. Data for the conditional probability of the magnitude of forces occurring in an accident for the predictive approach are described in Section 3.2. To use these data, a quantitative estimate of the force magnitude required for container failure is needed. The accident forces described in Section 3.2 are fire, impact, crush, and puncture. An estimate is needed of the container failure threshold for each force used in the analysis.

The approach generally used with the database presented in Section 3.2 is shown in Fig. 5-1. The four blocks outlined with heavy lines are representative of the process used to generate the database. Data for accident types were combined with data and assumptions for needed distributions, such as number of railcars derailing as a function of train speed. The two sets of "data" were used in engineering models incorporating container parameters to produce the figures and tables presented in Section 3.2. The usual approach for using the figures and tables is shown in the blocks of Fig. 5-1 outlined by light lines. The analyst defines accident scenarios (as described in Chapter 4) that produce accident forces (either impact, crush, puncture, or fire) that could threaten container integrity. For each force, the condition that would

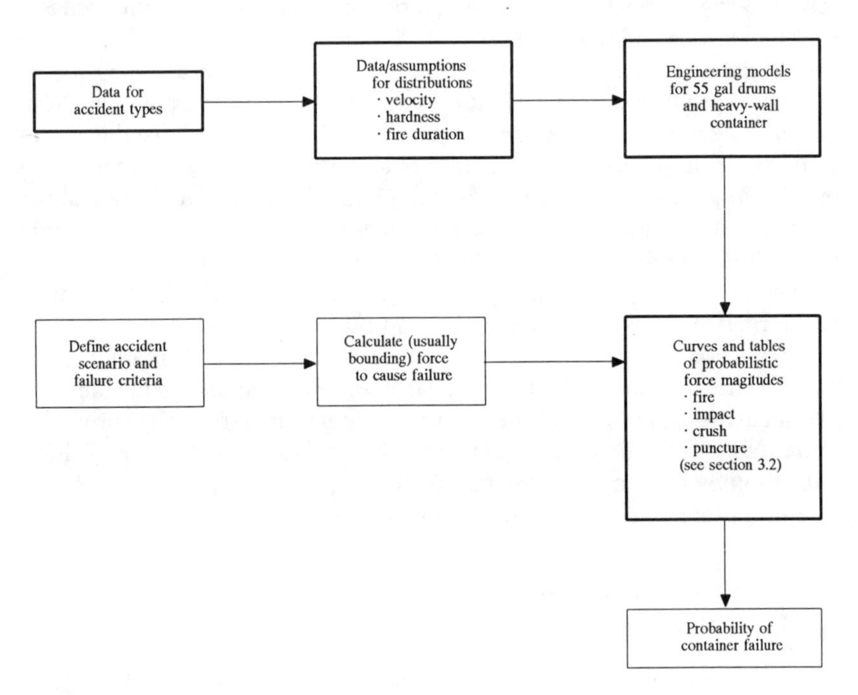

FIGURE 5-1. Approach for evaluating frequency of release using cumulative probability results.

lead to failure (e.g., the magnitude of a deflection, a stress, or a temperature) is defined. Using analytical tools appropriate to the project goals and resources, analyst determines the magnitude of the force required to produce the failure conditions. The force magnitude in the figures and tables is used to determine the probability of container failure. This approach is illustrated graphically in the top line of Fig. 2-3 and mathematically in Eq. 2-2.

The two blocks in the upper middle and the upper left-hand corner of Fig. 5-2 are conceptually the same as the two blocks in the upper middle and the upper left-hand corner of Fig. 5-1. In the approach illustrated in Fig. 5-2, a very detailed analysis of container failure modes was desired; therefore, response regions were defined that represented small leaks (e.g., seal failure) and large leaks (e.g., weld failure). For each force type, curves were generated analytically for a variety of parameters associated with accident conditions (e.g., strain as a function of impact velocity for a variety of angles at which the container might impact a surface). Each container response region results in one point on each container response curve. These points (force magnitude and accident parameters) are input into the "data" block to determine the container failure probability. This approach was used by Fischer et al. (1987).

The approach illustrated in Fig. 5-1, which uses the data presented in Section 3.2, is emphasized in this book. Other approaches using engineering models, such as that shown in Fig. 5-2, are possible. The use of engineering models is a characteristic of the bottom-up or predictive approach. The alternative is the top-down or historical approach using data similar to that presented in Section 3.3.

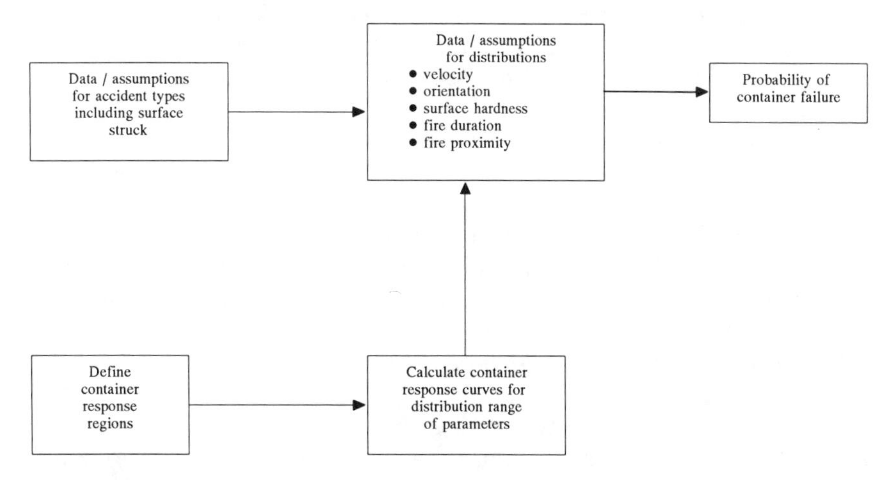

FIGURE 5-2. One alternative approach for evaluating frequency of release.

5.2 CONTAINER ANALYSIS APPROACHES

The primary alternatives for determining failure thresholds are: (1) values from the literature, (2) engineering judgment, (3) similarity analyses, (4) tests, and (5) direct calculation. Risk analyses in the literature may contain failure thresholds suitable for other risk analyses; an example is given next for a chlorine tank car. This is clearly the easiest way to obtain needed values. Results for chlorine (Andrews et al. 1980), gasoline (Rhoads et al. 1978), and propane (Geffen et al. 1980) frequently are used by other analysts.

Engineering judgment can be used to extrapolate from prior analyses and/or from the limited accident/test data available. This alternative may be the only practical one when resources are limited.

The concept of physical similarity is that two systems will behave similarly if ratios of physical quantities such as forces and velocities are the same. This concept led to the development of dimensionless parameters for heat transfer and fluid flow, such as the Reynolds and Nusselt numbers; this concept is also the basis for studying phenomena using scale models (Gröber et al. 1961). An application of this analysis technique is given here for a chlorine tank truck.

Tests usually are performed to show that a container meets regulatory criteria without failure. Therefore, existing test data usually would indicate that the failure forces are greater than those experienced in the test, but how much greater may be difficult to estimate. The testing approach for failure thresholds is: (1) estimate a force magnitude (e.g., drop height from impact testing) that would cause failure; (2) perform the test; and (3) raise or lower the test force to bracket the failure threshold. Clearly, depending on the estimating ability of the test director and the accuracy desired for the failure threshold, 5, 10, or more tests may be required. If containers are inexpensive, a practical testing approach may be used, as in the following illustration for a one-ton container.

Direct calculations fall into two categories: hand calculations and computer calculations. Hand calculations generally are limited to: simple geometries; linear, elastic structural analyses; and steady-state, one-dimensional heat transfer calculations. Computer analyses can be quite sophisticated. A complex heat transfer analysis can be relatively inexpensive compared to a complex structural analysis. For a complex container design, a complex structural analysis is potentially limited only by the financial resources of the project. An example of simple computer analysis is given in Section 5.3.4 for a tritium cylinder.

The approaches previously described have been presented in approximate order of their increasing demand on project resources. However, the accuracy of the analysis does not necessarily correspond to cost; for example, a good analysis from the literature is probably the least expensive and can also be the most accurate. Testing has the advantage of utilizing the real thing, whereas analytical models are, by necessity, mathematical approximations of the real world. On the other hand, with a reasonable analytic model, parameter studies can be easily performed, either to determine failure thresholds or to determine the effect of proposed mitigation strategies. As is the case for most aspects of risk analysis, the analyst must choose the approach that best suits the project objectives and resources. No one method is best for all or even most situations.

The objective of an engineering model is to show the relationship between the design features of interest and the accident forces of interest. If the scope of the risk analysis relates to the performance of only a portion of the design (e.g., the ability of the valves and the valve enclosure to resist accident forces), then the appropriate engineering model would cover these aspects in detail and the remainder of the container in as little detail as possible.

5.3 EXAMPLES OF PRACTICAL CONTAINER ANALYSES

The purpose of this discussion is to present results of analyses that can be used in later chapters, and very briefly to summarize the analytical approach used. Detailed mechanical and thermal calculations are beyond the scope of this book. Many hazardous material transportation situations involve bulk transport in containers that are relatively simple to model. Occasionally, the transportation geometry is complex, such as the transport of munitions, and the thermal and mechanical analyses can be very complex. For example, a three-dimensional, nonlinear (materials and geometry), elastic-plastic analysis was desired for an armored, insulated, reusable overpackage for chemical munitions (Pomerening and Cox 1985).

5.3.1 Chlorine Tank Car

Chlorine must be transported in designated tank cars, typically the 105A500W railcar shown in Fig. 4-4. Failure thresholds for this car were determined by Andrews et al. (1980). A brief overview of the analytical techniques used to determine the failure thresholds is presented here. Selected results given in Table 5-1 will be utilized in Chapter 8.

The velocity producing failure in a head impact was found by equating the initial kinetic energy to the work required for inelastic expansion of the tank diameter. The failure criterion of a 5% diametrical expansion was based on pressure vessel burst tests. The side velocity for failure was found by equating the work required to compress the chlorine to the rupture (ultimate stress) pressure with the kinetic energy. (Defective tank car failure thresholds also were developed, but, as shown in Chapter 8, reasonable values for the occurrence of defective transportation containers result in insignificant effects for risk analyses.)

Puncture failure probabilities in Section 3.2 are determined from an equivalent steel thickness. Note that the head value in Table 5-1 includes a 0.5-in.-thick head shield. The equivalent thickness of the shield plus head is 1.16 in. (Andrews et al. 1980).

To make the crush calculation applicable to all crush configurations, the stiffening effects of the ends were neglected. Failure was assumed to occur when the internal tank volume was reduced by shell deformation such that the hydrostatic pressure exceeded the tank rupture pressure. Equations for deflection of a cylinder under uniform loading were used along with pressure-volume relationships for the gas phase. The liquid was considered incompressible.

Thermal calculations were performed for three release mechanisms. The first mechanism is initiated when the tank reaches the relief valve pressure of 375 psig. If the relief valve opens, chlorine will be released to control the pressure rise. If the relief valve fails to open, Andrews et al. (1980) assumed that at the relief pressure the liquid volume occupies the car volume com-

TABLE 5-1. Failure thresholds for the 90-ton 105A500 railcar containing chlorine

(5-1a)

	Failure Threshold	
Failure force type	Shell	Head
Impact	18 mph	32 mph
Puncture (steel equivalent thickness)	0.787 in.	1.16 in.
Crush	134,000 lb distributed along length	

(5-1b)

	Failure Threshold	
Fire scenario	Insulated	10% Insulation loss
Internal pressure is 375 psig[a]	100 min	35 min
Upright tank fails as level falls[b]	290 min	100 min
Overturned car fails owing to insufficient relief rate of liquid[c]	164 min	55 min

Compiled from: Andrews et al. 1980.

[a]Tank car will soon rupture if relief valve fails, producing an instantaneous release. If valve operates, a continuous release results. [b]Tests show that at about 50% full, weakened wall collapses owing to lack of cooling from liquid. [c]Note that vapor release carries out more energy than does liquid.

pletely so that the pressure then will rise very rapidly to produce failure by overpressure.

Full-scale tests have shown that a tank car in a fire can fail when the liquid level drops to one-half full because the liquid is no longer removing heat from the top part of the tank. Thus, the second release mechanism is failure of the tank due to a high wall temperature.

If the tank car is overturned, predominantly liquid chlorine will be discharged from the relief valve until the liquid level falls to the elevation of the valve. If the heating rate is high enough, the tank pressure may continue to rise. Andrews et al. (1980) conservatively assumed tank failure at 500 psi rather than the 1250 psi design pressure for this situation. Time-dependent energy and mass balances were solved numerically for the liquid and vapor phases. Analyses were performed for the case of no damage to the insulation and for the case of 10% insulation removal. The latter might occur in a raking collision along the length of the tank shell.

5.3.2 Chlorine Tank Truck

Chlorine typically is transported in a special MC 331 tank truck that has a 4-in.-thick minimum insulation thickness and a minimum wall thickness of

TABLE 5-2. Failure thresholds for the 20-ton MC-331 tank truck containing chlorine

(5-2a)

	Failure Threshold	
Failure force type	Shell	Head
Impact	32 mph	56 mph
Puncture (steel equivalent thickness)	0.625 in.	0.625 in.
Crush	84,000 lb distributed along length	

(5-2b)

	Failure Threshold	
Fire scenario	Insulated	10% Insulation loss
Internal pressure is 375 psig[a]	50 min	18 min
Upright tank fails as level falls[b]	145 min	50 min
Overturned car fails owing to insufficient relief rate of liquid[c]	82 min	28 min

[a]Tank truck will soon rupture if relief valve fails, producing an instantaneous release. If valve operates, a continuous release results. [b]Tests show that at about 50% full, weakened wall collapses owing to a lack of liquid chlorine cooling. [c]Note that vapor release carries out more energy than does liquid.

0.625 in. (Title 49 Code of Federal Regulations). The analyses of the chlorine railcar can be used as the basis for establishing failure thresholds using physical similarity concepts. The impact velocity determined in Section 5.3.1 can be shown to be proportional to the square root of (T/LD), the crush force is proportional to T^2L/D, and the fire durations are proportional to the volume (ΠD^2L) divided by surface area (πDL), that is, D. In these relationships, T is the tank thickness, L the length, and D the diameter. The values from this analysis are presented in Table 5-2.

5.3.3 Ton Container

The ton container shown in Fig. 5-3 is used for some materials, such as anhydrous hydrofluoric acid and chlorine. The DOT specification class is 106A and 110A (The Chlorine Institute 1980). The failure thresholds for the ton container are given in Table 5-3. The impact failure threshold was established from drop tests; the velocity was determined by equating kinetic energy with potential energy. No distinction was found for side impact and impact at a 45-degree angle; therefore, the end impact failure value was chosen equal to the side and 45-degree value.

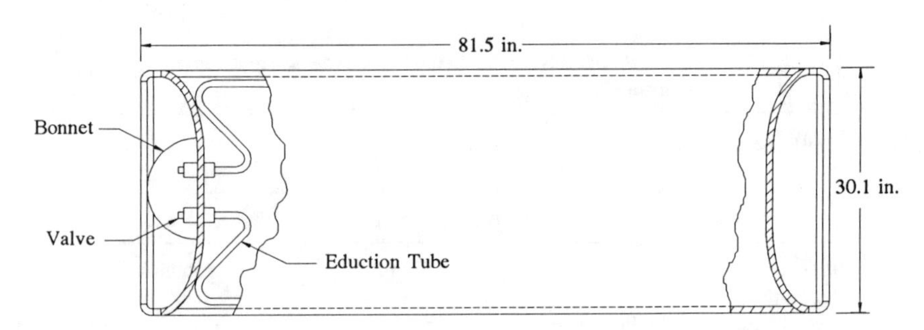

FIGURE 5-3. Ton container. Adapted from: The Chlorine Institute 1980.

TABLE 5-3. Failure thresholds for a ton container

Failure force type	Failure threshold
Impact	35 mph
Crush	245,000 lb
Puncture	7/16 in. equivalent thickness
Fire	30 min (overpressure); 4 min (if equipped with fusible plugs)

The crush value was obtained by equating the work of producing a 6-in. deformation along the side with the kinetic energy from the drop test. The value shown is about twice that obtained from a calculation that ignored end stiffening. The fire model applied a simple finite difference scheme to the standard heat balance equation to determine the container wall temperature and the liquid temperature; the ideal gas law was used to estimate the quasi-static pressure. The rupture pressure at the elevated temperature was taken from time-to-rupture curves. When the container is equipped with fusible plugs, the time to release is short; thus, failure at high pressure is avoided.

5.3.4 Tritiated Water Container

The container shown in Fig. 5-4 was proposed for transport of tritiated water, a radioactive substance. The outer container is a 55-gal steel drum, either a DOT Specification 17C drum or a DOT Specification 6C drum. The function of the drum is to hold in place and protect insulation surrounding the inner container.

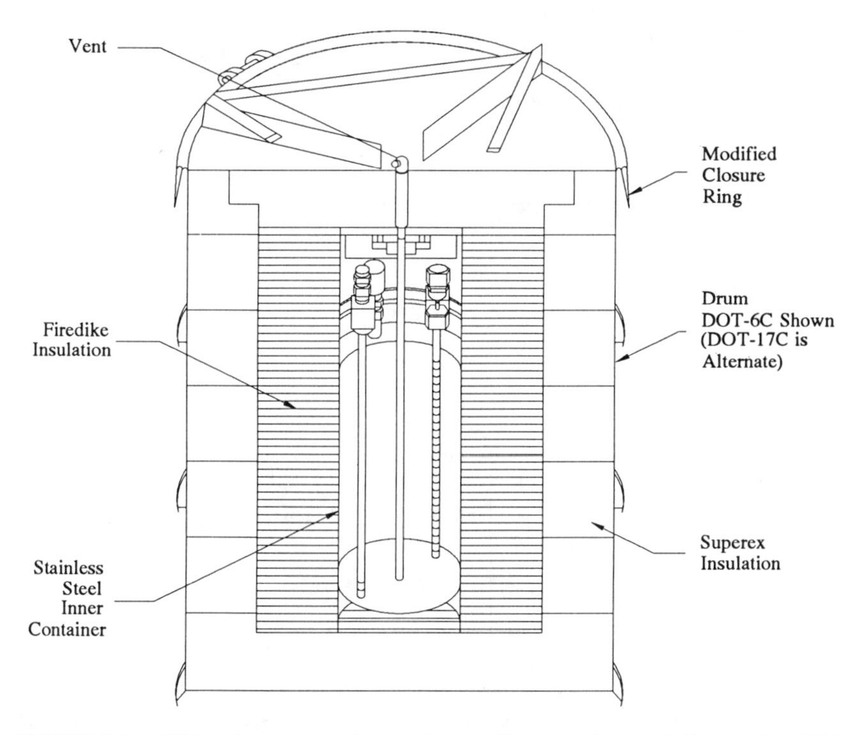

FIGURE 5-4. Tritiated water container. Source: Monsanto Research Corporation 1987.

The outer layer of insulation is fixed within the drum and is approximately 3.5 in. thick. The inner layer of insulation is removable and is approximately 4 in. thick. The insulation protects the inner container from thermal forces and provides some impact and puncture force protection.

The inner container, shown in more detail in Fig. 5-5, is a cylindrical vessel nominally 6⅝-in. o.d by 23⅞-in. overall height; it has a minimum wall

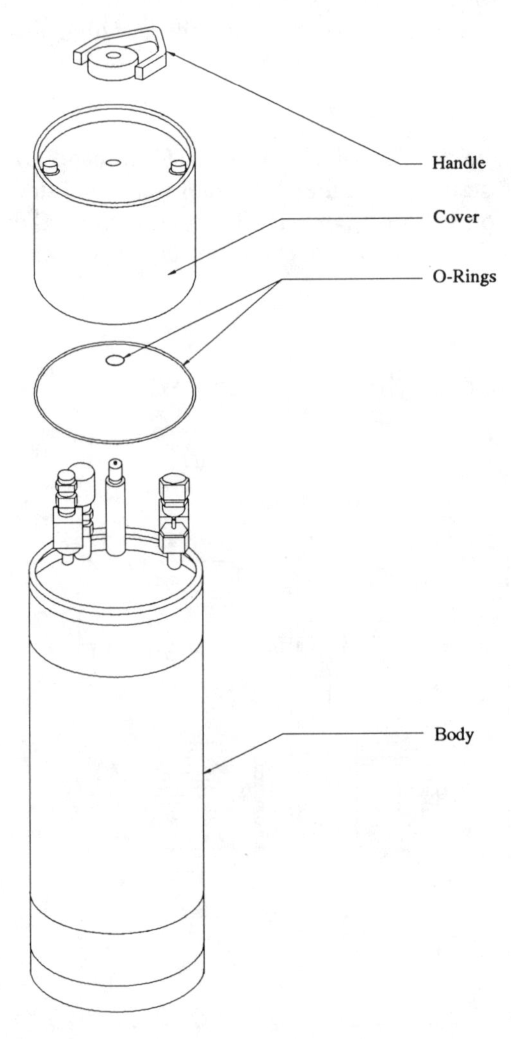

FIGURE 5-5. Inner container—AL-M1, configuration 5. Source: Monsanto Research Corporation 1987.

thickness of 0.12 in. The vessel is made of 316 stainless steel and has a top cap sealed with O-rings. The function of the top cap is to cover bellows valves, fittings, and a pressure transducer. The vessel is loaded with molecular sieve pellets or other material for sorption of tritiated water (Monsanto Research Corporation 1987).

The acceleration of the inner vessel due to 30-ft drops of the entire container onto an unyielding surface was analyzed for side, top, bottom, and top corner impact orientations. The structural insulation provides a cubic elastic load-deflection function for transmitting loads to the inner vessel. The inner vessel was modeled with a conventional three-dimensional finite element software package. The maximum acceleration transmitted to the inner vessel resulted from a side drop. The stress contours that result from the 30-ft side drop are shown in Fig. 5-6. Impact loading on the valves mounted on the inner vessel head was determined to cause the head to fail (Bales et al. 1990). This quasi-static analysis represents the first level of dynamic numerical structural analysis. The next level would consider the time-dependent loading of the vessel. The quasi-static approach is conservative and the less expensive of these methods; therefore, it was the most cost-effective approach for the requirements of this analysis.

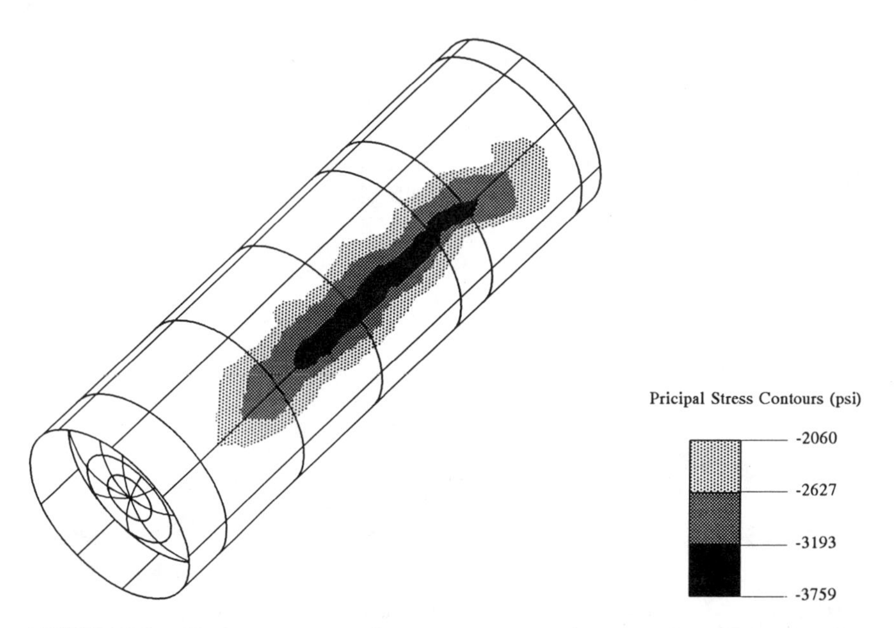

FIGURE 5-6. Maximum (compressive) stress contours that result from 30-ft side drop. Source: Bales et al. 1990.

For thermal analysis, the container was modeled in two dimensions as shown in Fig. 5-7. The response of the container to a 30-min fire producing a radiation environment of 1475°F (800°C) over the entire package was determined. The radioactive decay of the vessel contents was modeled as a uniform 3.3-watt heat source. The time history (in °R) of selected nodes from Fig. 5-7 is shown in Fig. 5-8. All node temperatures for the vessel containment boundary are at least 580°F below their rated service limit, and no significant thermal stresses are developed. The fire duration required for

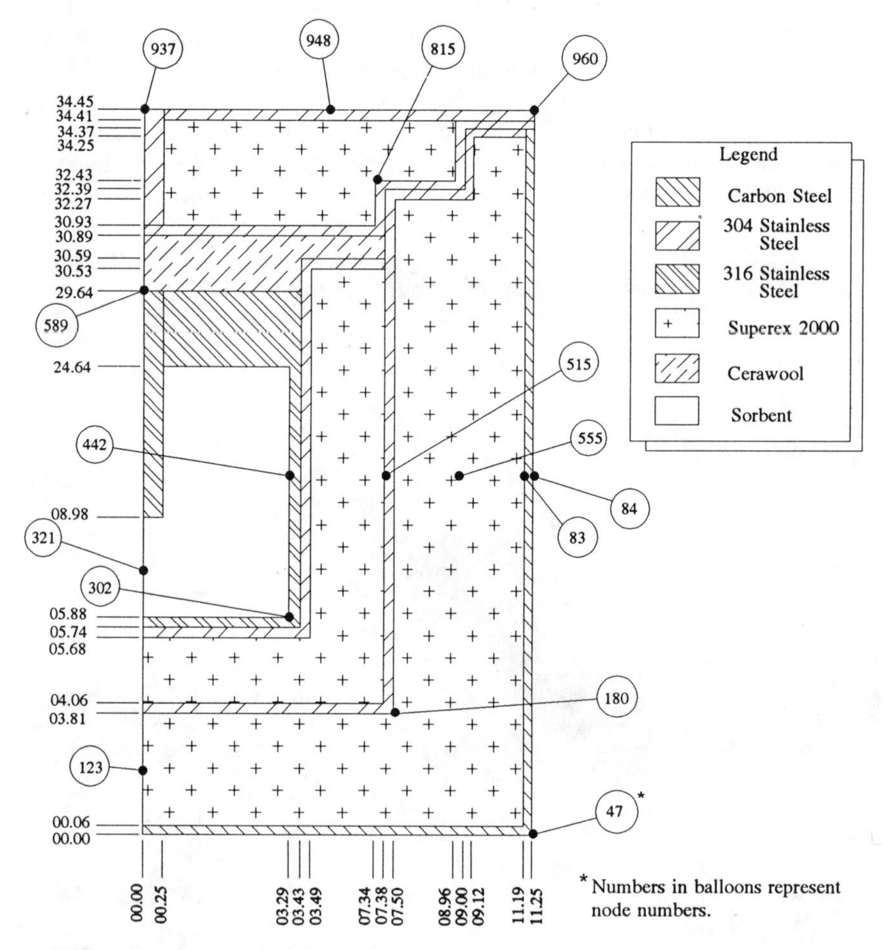

FIGURE 5-7. Analytical axis-symmetric thermal model of the AL-M1 Shipping Container. Source: Bales et al. 1990.

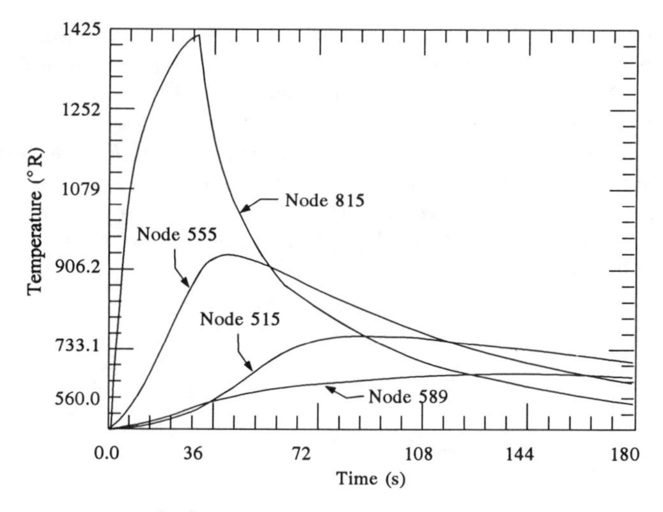

FIGURE 5-8. Container temperature history plot. Source: Bales et al. 1990.

container failure is significantly greater than 30 min. Because this particular analysis was conducted to show compliance with a 30-min fire regulatory requirement, a fire of longer duration was not analyzed.

References

Note: The reports of U.S. government agencies, their laboratories, and contractors cited here are available from the National Technical Information Service, Springfield, Virginia 22161, USA.

Andrews, W. B., et al. March 1980. *An Assessment of the Risk of Transporting Liquid Chlorine by Rail.* PNL-3376. Pacific Northwest Laboratory.

Bales, P. A., R. W. Emmett, A. K. Langston, and J. M. Singleton. 1990. *Supplemental Structural and Thermal Analysis of the Model Al-M1 Configuration 5 Container.* Oak Ridge, Tennessee: H&R Technical Associates, Inc.

Fischer, L. E., et al. 1987. *Shipping Container Response to Severe Highway and Railway Accident Conditions.* NUREG/CR-4829. Lawrence Livermore National Laboratory.

Geffen, C. A., et al. March 1980. *An Assessment of the Risk of Transporting Propane by Truck and Train.* PNL-3308. Pacific Northwest Laboratory.

Gröber H., S. Erk, and U. Griyull. 1961. *Fundamentals of Heat Transfer.* New York: McGraw-Hill Book Co., Inc.

Monsanto Research Corporation. 1987. *Draft for Review, Safety Analysis Report (SARP): Model Al-M1 Nuclear Packaging.* MLM-3446. U.S. Department of Energy.

Pomerening, D. J., and P. A. Cox. July 1985. *Impact Analysis of the CAMPACT Shipping Container.* Southwest Research Institute Final Report for Project No. 06-8461-002. AMXTH-CD-TR-86056 (available from the Defense Technical Information Center).

Rhoads, R. E., et al. November 1978. *An Assessment of the Risk of Transporting Gasoline by Truck.* PNL-2133. Pacific Northwest Laboratory.

The Chlorine Institute. June 1980. *Cylinder and Ton Container Procedure for Chlorine Packaging.* Chlorine Institute Pamphlet 17, Edition 1. Washington, D.C.: The Chlorine Institute.

6

Consequence Analysis

The consequence of an accident scenario can be determined by a three-step procedure: (1) definition of the source term, that is, the release amount and mode of release; (2) exposure assessment, that is, the extent to which people are exposed to the source term; and (3) assessment of the health effect. The health effect is the consequence of interest. Risk is defined as frequency and consequence in Chapter 1; therefore, all three steps are considered as the consequence portion of the risk analysis. A further breakdown of the consequence calculation procedure is shown in Fig. 6-1.

The consequence portion of the risk analysis begins with the release of hazardous material from the container, either because the container has failed from accident forces or because a relief valve has opened, perhaps owing to high pressure caused by a fire. Nonaccident releases usually are due to a misoperation such as a valve not closed tightly, and those release rates and amounts are not addressed quantitatively in this chapter (see Sections 3.3 and 3.4). However, the exposure and health assessments of this chapter apply to nonaccident releases.

Accident forces initiate the accident scenarios described in Chapter 4 that result in the container failure or release modes. The accident scenario description includes the container failure/release mode. The hazardous material discharge may be essentially instantaneous, very slow, or somewhere in between. The released material may form a vapor cloud, a pool of liquid, or a momentum jet. If the material is flammable, immediate ignition may occur, producing an explosion or a fire. If ignition is delayed, the concern is that the vapor cloud may grow even larger before combustion commences, with even larger potential consequences. If the material is toxic, the concern is that people may be exposed as the released material is dispersed downwind. Peo-

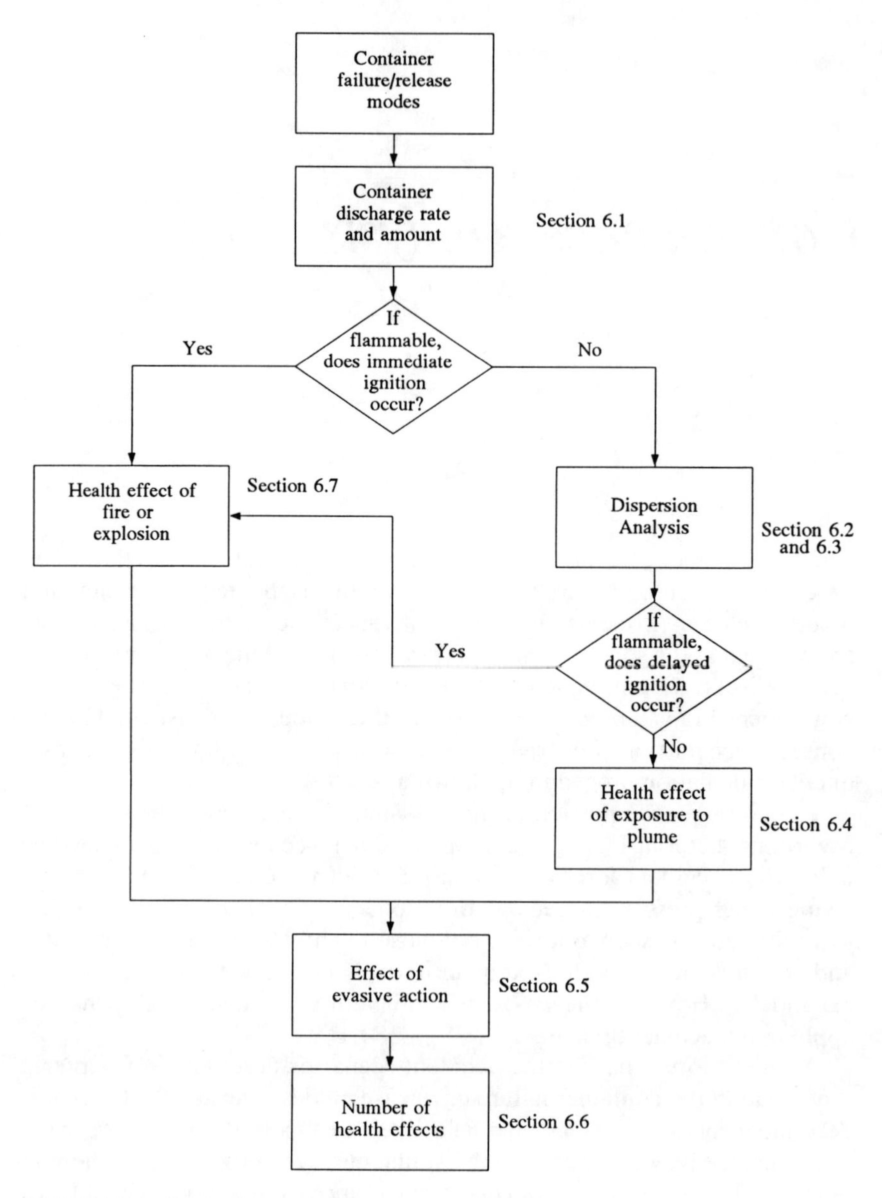

FIGURE 6-1. Logic diagram for consequence analysis.

ple may be exposed to the flammable and possibly toxic cloud. Some people in the affected downwind area may seek shelter or escape from the affected area. For those persons exposed, the health effects are estimated.

The physical processes involved in the release and dispersion of hazardous materials following accidental releases are very complex and in some cases not well understood. Entire books are written on some of these topics; therefore, the level of detail here must be very selective. The major considerations at each stage of the procedure are presented and toxic-by-inhalation materials emphasized.

6.1 RELEASE RATE AND AMOUNT

The nature of the release depends on the location of the hole, the size of the hole, and the conditions in the container. Some of these considerations are illustrated in Fig. 6-2.

6.1.1 Discharge from Massive Failure

Release type 1 is from a sudden, massive failure of the container. If the material is a liquid whose boiling point is less than the ambient temperature, the sudden decrease in pressure results in the vaporization of a fraction of the liquid. A substantial amount of liquid will be entrained in the vapor cloud and carried from the container with the vapor.

A boiling-liquid, expanding-vapor explosion (BLEVE) may result if the depressurization is sufficiently rapid (McDevitt et al. 1990). Parts of the container may be propelled from the explosion. The exact mechanisms and conditions for a BLEVE are not well understood (Leslie and Birk 1991). A BLEVE may result from several mechanisms including mechanical damage, particularly collisions; mechanical failure, such as weld failure; overfilling or

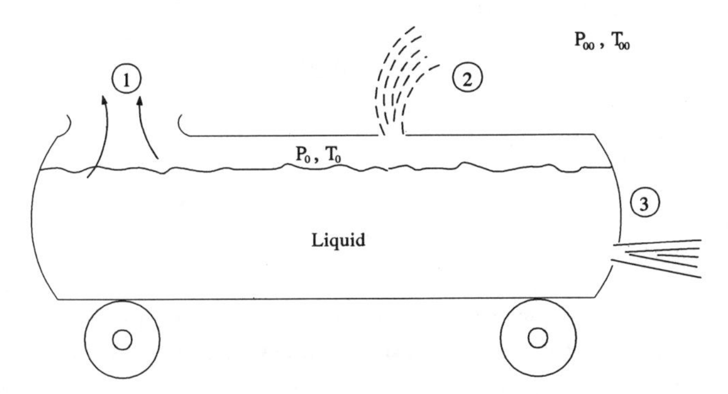

FIGURE 6-2. Illustration of release types.

overheating without a relief mechanism; or a runaway reaction. However, the typical BLEVE is caused by a fire external to a container of liquid (Prugh 1991). In the case of an external fire, the tensile strength of the container material, particularly in the unwetted portion, is reduced by the elevated wall temperature. The BLEVE is initiated because the container material is not able to withstand the internal pressure even though the relief valve(s), if present, is open. Historically most BLEVEs involve flammable liquids; therefore, the released material is ignited by the fire that caused the BLEVE. A fireball results only if the material is flammable.

The prediction of the extent of the release of vapor and entrained liquid is very uncertain for large-sized failures in containers of liquefied gases. If the pressure is relieved slowly, the extent of liquid vaporization is determined from thermodynamic properties. For chlorine, about 20% would flash to vapor, and the remainder would evaporate. Tests and accident experience show that for large-sized failures, massive quantities of liquid are ejected from the container and become airborne. Tests described in Section 6.2 show that large amounts of aerosolization will occur for releases of sub-cooled liquids. Massive failures from mechanical forces in transportation accidents are likely to be on the side or the bottom of containers; therefore, a complete release, essentially instantaneous, is modeled unless data to the contrary exist for a specific commodity. Massive failures from thermal forces also are appropriately modeled as complete and instantaneous.

6.1.2 Gas Discharge

Release type 2 is a gas discharge from the vapor space of the container, and the release rate is described by standard equations (Sakiadis 1984). If the container pressure is about twice the ambient pressure, then the gas discharge is limited to choked flow; that is, the flow velocity is equal to the speed of sound in the gas. The criterion for choked flow is:

$$P/P_a \geq [(\gamma + 1)/2]^{\gamma/(\gamma - 1)} \qquad (6\text{-}1)$$

where:

P = absolute tank pressure (N/m^2)
P_a = absolute ambient pressure (N/m^2)
γ = gas specific heat ratio

For a typical gas, γ is about 1.5, and the critical pressure ratio from Eq. 6-1 is 1.95, or about twice the absolute ambient pressure. The flow rate for choked (sonic) flow of an ideal gas is:

$$Q = C_d AP[(\gamma M/RT)(2/(\gamma + 1))^{(\gamma + 1)/(\gamma - 1)}]^{1/2} \qquad (6\text{-}2)$$

where:

Q = discharge rate (kg/s)
C_d = orifice discharge coefficient
A = flow area (m^2)
M = gas molecular weight (kg/kg-mole)
R = gas constant (8.31×10^3 J/kg-mole/K)
T = absolute gas temperature in container (K)

The flow will remain constant as long as the pressure ratio given by Eq. 6-1 is satisfied.

The subsonic discharge rate for an ideal gas is:

$$Q = C_d A\{2\rho_g P(\gamma/(\gamma - 1))[(P_a/P)^{2/\gamma} - (P_a/P)^{(\gamma+1)/\gamma}]\}^{1/2} \tag{6-3}$$

where ρ_g = container gas density (kg/m^3). The subsonic gas discharge rate will decrease as the container pressure drops.

Most gas dispersion models use a constant gas discharge rate as input; therefore, generally the initial, maximum value from either Eq. 6-2 or Eq. 6-3 is used. The associated discharge time is the initial gas mass divided by the initial value of the release rate.

If the discharge area is large (perhaps five times that of a typical gas relief valve), the gas discharge even from the top of the container holding a low-boiling-point liquid will start to exhibit characteristics of two-phase flow. If the container failure location is on the side of the tank near the liquid phase level, the flow is likely to exhibit the two-phase characteristics described in the next section.

6.1.3 Liquid Discharge

Discharge of a liquid is described by the Bernoulli equation:

$$Q = C_d A[2(P - P_a)\rho_f + 2gh]^{1/2} \tag{6-4}$$

where:

p_f = liquid density (kg/m^3)
h = height of liquid above discharge point (m)
g = gravitational acceleration (9.8 m/s)

(In transportation accident situations for liquids or gases under pressure, the gravity head usually can be ignored.) The discharge coefficient is approximately 0.6 (Fauske and Epstein 1988). If the liquid in the container is pressurized and the boiling point is above the ambient temperature, the discharge rate is approximated as a constant (Fauske and Epstein 1988) by:

$$Q = C_d A[2(P - P_v(T_0))\rho_f]^{1/2} \quad (6\text{-}5)$$

where $P_v(T_0)$ is the vapor pressure at the container temperature (N/m^2). The approximate discharge time is the initial mass divided by Q. Because $P_v(T_0)$ is greater than P_a for the liquid under consideration, the pressure difference in Eq. 6-5 is less than that in Eq. 6-4, and the flow rate computed from Eq. 6-5 is less than that computed from Eq. 6-4. The average rate computed from Eq. 6-5 over the discharge time in the model produces a reasonable approximation to experimental data (Fauske and Epstein 1988).

6.2 TRANSITION FROM SUPERHEATED LIQUID DISCHARGE TO DISPERSION

If the liquid is superheated at ambient conditions, some of it will flash to vapor when discharged. The fraction of the superheated liquid that will flash is predicted from thermodynamics as:

$$f = C_p(T_s - T_b)/h_{fg} \quad (6\text{-}6)$$

where:

T_s = storage temperature (K)
T_b = boiling temperature at ambient pressure (K)
h_{fg} = heat of vaporization (J/kg)
C_p = specific heat at constant pressure of the liquid (J/kg/K)

The remaining liquid will have been cooled to its boiling point, and some liquid will remain in the cloud as a finely dispersed liquid, or an aerosol. Some aerosol will evaporate owing to heat transfer with the entrained air and possible exothermic reactions with moist air, and some will "rain" from the cloud, forming a pool on the ground. The subsequent evaporation of volatile liquids owing to heat transfer from the ground may be so rapid that for all practical purposes the rainout can be neglected. The net result is that the use of Eq. 6-6 generally will significantly underestimate the mass of material in the cloud.

The aerosol component has a significant effect on the subsequent cloud dispersion. Because of this component, the cloud mass is denser; evaporating aerosol will reduce the cloud temperature and hence increase the cloud density; and the cold cloud may cause entrained atmospheric moisture to condense, increasing the density further. Thus, to initialize cloud dispersion models, one first must determine the amount of aerosol entrained in the cloud.

Some chemicals form complex molecules (e.g., $(HF)_n$ where n values range from 1 to 8), which increases the vapor density. Reactions with water vapor are not unusual. For example, water vapor and HF react to form a liquid phase, but the reaction is exothermic, so that the temperature tends to rise and the cloud buoyancy to increase. As the $(HF)_n$ is diluted in dry air by dispersion, endothermic dissociation occurs, also producing opposing effects on cloud density (Clough et al. 1987).

The transition from the discharge of a flashing liquid to the formation of a cloud that contains two-phase material is not well understood. To help analysts understand the phenomenology, tests and first-principles analyses were initiated in the late 1980s. The tests indicate little aerosolization at low levels of superheat, $T_s - T_b$. As the amount of superheat increases to about 25°C, the liquid being discharged starts to break up, that is, changes from a continuous stream of liquid to a stream of liquid drops. Further increases in superheat produce rapid increases in the extent of aerosolization. In 1990, tests with chlorine released with as little as 10°C of superheat resulted in vaporization or aerosolization of 75% of the liquid released; yet Eq. 6-6 indicates that only 8.6% would flash. At about 55°C of superheat, 100% of the released chlorine is vaporized or aerosolized. The analytical model predicts the shape of the curve of vaporization/aerosolization versus superheat, but not the specific values just given (Johnson 1991). These tests indicate that releases of a liquid with a boiling point more than 25°C below ambient temperature should be modeled by assuming that all of the discharged liquid immediately enters the cloud as vapor or aerosol.

At this point in the analysis, the composition of the material to be dispersed is assumed known, that is, a specific mixture of vapor and aerosol. The next section will show how the effective volume of the cloud at the start of dispersion can be considered.

6.3 DISPERSION MODELS

The atmospheric dispersion of vapor clouds depends on three main factors: the characteristics of the released material, the meteorological conditions, and the terrain. The physical conditions causing and controlling the release determine the mass release characteristics as described in the previous sections; the downwind concentration is proportional to the mass release rate or the total mass released. The density of the material (including the presence of aerosols) determines whether the cloud is heavier than air and tends to sink, is neutrally buoyant, or is positively buoyant and tends to rise. The initial movement of the cloud also will be affected by any initial momentum. The duration of the release relative to the point of exposure determines whether the release appears to be instantaneous or continuous. Last, both the geometric size of the release before downwind dispersion begins and the height of the release affect the results.

Of the source characteristics just given, the initial density and the release duration affect the first choices of the analyst:

1. Either the dense gas model or the Gaussian model.
2. Either a continuous formulation or a puff formulation.

Given these choices, the meteorological conditions and the terrain have an effect on the parameter values of the model formulation as described in the following paragraphs and in the next section.

The Gaussian model is important historically and is still by far the most commonly used. Gaussian modeling information will be presented quantitatively here because of the extensive use of Gaussian models, including their use for dense gas dispersion. Full details, however, are found in the references (e.g., Gifford 1968, Turner 1970, and Hanna et al. 1982).

Dense gas models are complex; even relatively simple versions are performed only on computers. Further, dense gas models are rapidly evolving; therefore, the model description in this book will be qualitative. Two of the many reviews of the dense gas models are contained in Hanna and Drivas (1987) and Hanna et al. (1982).

The primary meteorological parameters are the wind speed and the atmospheric stability. The relative humidity is important if the vapor cloud constituents react with water vapor. The wind speed usually is measured at 10 m above the surface. Atmospheric stability usually is defined by the vertical temperature gradient: a neutral stability exhibits a temperature decrease of about 1°C for each 100 m above the surface. An unstable temperature profile has a more negative gradient, which produces additional turbulence, and a stable temperature profile has a less negative gradient, which suppresses turbulence (Hanna et al. 1982). In the absence of site-specific stability measurements, the Pasquill scheme shown in Table 6-1 is the most widely used reference (Pasquill 1961 as modified by Turner 1964: both cited by Gifford 1976). Daytime insolation can be classified as strong when the solar altitude is greater than 60° with clear skies and as slight when the solar altitude is 15° to 35° with clear skies. Insolation that would be strong with clear skies is reduced to moderate with 5/8 to 5/8 cloud cover if

TABLE 6-1. Meteorological conditions defining Pasquill stability types

Surface wind speed, m/sec	Daytime Insolation			Nighttime Conditions	
	Strong	Moderate	Slight	Thin overcast or $\geq 4/8$ cloudiness[b]	$\leq 3/8$ cloudiness
<2	A	A-B	B		
2	A-B	B	C	E	F
4	B	B-C	C	D	E
6	C	C-D	D	D	D
>6	C	D	D	D	D

Source: Gifford 1976.

Notes: A: extremely unstable conditions. B: moderately unstable conditions. C: slightly unstable conditions. D: neutral conditions.[a] E: slightly stable conditions. F: moderately stable conditions.
[a]Applicable to heavy overcast day or night. [b]The degree of cloudiness is defined as that fraction of the sky above the local apparent horizon that is covered by clouds.

the cloud height is intermediate and slight if the cloud height is low (Turner 1970). Hanna et al. (1982) recommend the use of the Richardson number to define stability because it is a measure of both the temperature gradient and mechanical shear from wind. The Richardson number frequently is used in dense gas models to determine when to change to a neutrally buoyant model.

Terrain conditions affecting dispersion include the effects of a rough surface (buildings or trees) or a smooth surface (grass). Flow around buildings has been reviewed by Hosker (1982).

6.3.1 Gaussian Model

The primary assumption of the Gaussian model is that the atmospheric turbulence is random, and therefore the concentration distribution is Gaussian in the vertical and the horizontal (perpendicular to the downwind) directions. The model produces results that agree with data as well as more sophisticated models do. Because the Gaussian model is reasonably accurate and is easy to use, it is probably the most extensively used dispersion model. Additional assumptions are that the cloud constituents are inert and that the plume is reflected by the ground rather than being absorbed. The ground roughness can affect the turbulence, however.

The Gaussian formulas given next (Gifford 1968) are for the practical situation in which the person(s) exposed is at ground level. The formula for continuous release is:

$$\chi = (Q/\pi \sigma_y \sigma_z u)\exp\{-[(y^2/2\sigma_y{}^2) + (h^2/2\sigma_z{}^2)]\} \tag{6-7}$$

and the formula for a puff or instantaneous release is:

$$\chi = [Q/(2^{1/2}\pi^{3/2}\sigma_x\sigma_y\sigma_z)]\exp\{-[((x-ut)^2/2\sigma_x^2) + (y^2/2\sigma_y^2)+(h^2/2\sigma_z^2)]\} \quad (6\text{-}8)$$

where:

χ = concentration of released material (g/m^3)
Q = material release rate (g/s) for continuous release or total material release (g) for puff releases
σ_x, σ_y, σ_z = standard deviations of the concentration distribution in the x, y, and z directions, respectively
y = horizontal distance perpendicular to the plume centerline (m)
z = vertical distance perpendicular to the plume centerline (m)
x = distance downwind (m)
h = effective release height from the ground (m)
u = wind speed (m/s)
t = time since release (s)

A major difficulty is the proper selection of the diffusion parameters σ_y and σ_z. Figure 6-3 shows the Pasquill-Gifford curves developed from data for ground-level releases on relatively smooth, level terrain. Table 6-2 shows the formulas recommended by Briggs (1973) (as cited in Hanna et al. 1982) for open country, which are based on the Pasquill-Gifford curves in the near field and data for elevated releases in the far field. The urban values in

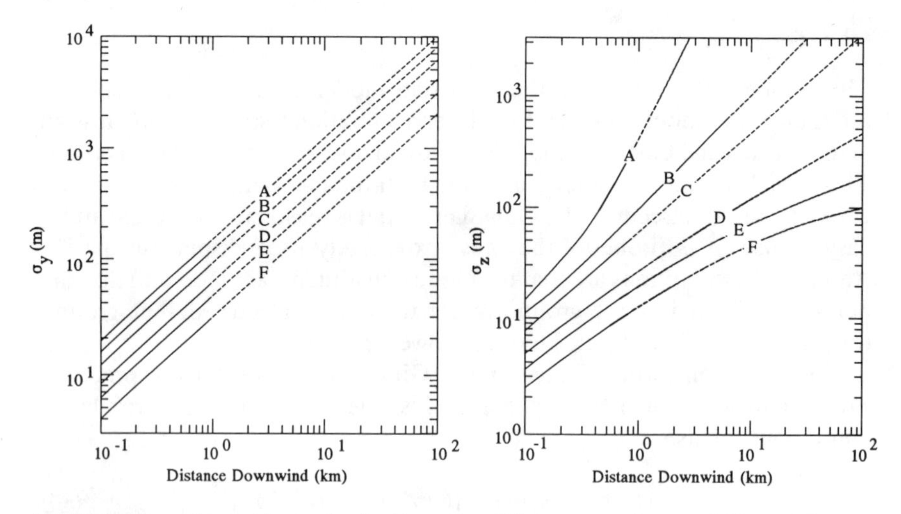

FIGURE 6-3. Curves of σ_y and σ_z for Pasquill stability types. Source: Gifford 1976.

TABLE 6-2. Formulas recommended by Briggs for $\sigma_y(x)$ and $\sigma_z(x)$ ($10^2 < x < 10^4$m)

Pasquill stability type	σ_y, m	σ_z, m
	Open-Country Conditions	
A	$0.22 \times (1 + 0.0001x)^{-1/2}$	$0.20x$
B	$0.16 \times (1 + 0.0001x)^{-1/2}$	$0.12x$
C	$0.11 \times (1 + 0.0001x)^{-1/2}$	$0.08 \times (1 + 0.0002x)^{-1/2}$
D	$0.08 \times (1 + 0.0001x)^{-1/2}$	$0.06 \times (1 + 0.0015x)^{-1/2}$
E	$0.06 \times (1 + 0.0001x)^{-1/2}$	$0.03 \times (1 + 0.0003x)^{-1}$
F	$0.04 \times (1 + 0.0001x)^{-1/2}$	$0.016 \times (1 + 0.0003x)^{-1}$
	Urban Conditions	
A-B	$0.32 \times (1 + 0.0004x)^{-1/2}$	$0.24 \times (1 + 0.001x)^{1/2}$
C	$0.22 \times (1 + 0.0004x)^{-1/2}$	$0.20x$
D	$0.16 \times (1 + 0.0004x)^{-1/2}$	$0.14 \times (1 + 0.0003x)^{-1/2}$
E-F	$0.11 \times (1 + 0.0004x)^{-1/2}$	$0.08 \times (1 + 0.00015x)^{-1/2}$

Source: Hanna et al. 1982.

Table 6-2 are based on data over St. Louis. Another way to account for urban σ_y and σ_z is to increase the stability class one level; for example, for C urban conditions use B rural values (Hanna et al. 1982). Gifford's (1976) review of data for σ_z near highways showed very little organization by stability class, but values were significantly increased over the curves in Fig. 6-3 for downwind distances up to 200 to 400 m. For most analyses of toxic plumes (10-30-min exposures) the use of Fig. 6-3 for near-ground releases associated with transportation accidents should be appropriate for rural highways. For urban highways, Briggs's values in Table 6-2 generally would be appropriate. Figure 6-3 and Table 6-2 are based on an averaging time of 10 min. The averaging time smooths out plume fluctuations owing to variations in the wind velocity. The longer the averaging time is, the broader the apparent plume width. If the plume is flammable, an instantaneous concentration is needed, and the estimated effect of a zero averaging time on the horizontal dispersion parameter σ_v is to reduce it by a factor of two (TNO 1992a).

The lower limit of validity for the values of σ, and hence the Gaussian model, is 100 m, as seen in Fig. 6-3 and Table 6-2. Beyond 800 m, the values in Fig. 6-3 are extrapolations, but the values in Table 6-2 are based on data to 20,000 m.

The preceding discussion of σ_y and σ_z is for continuous plumes. The data for estimating the corresponding values for puff releases are sparse. Hanna et al. (1982) recommend a formulation for σ_y^2 that is proportional to the cube of the puff travel time, and $\sigma_z = 0.3$ times the mixing layer height, but this application is not straightforward. Usually it is assumed that $\sigma_x = \sigma_y$, and continuous plume data are used for σ_y and σ_z.

6.3.2 Dense Gas Models

Development of dense gas models began about 20 years ago when tests using dense gases showed that the vertical standard deviation, σ_z, was about one-fourth the value for a neutrally-buoyant gas, and the horizontal standard deviation, σ_y, was about four times the value for a neutrally-buoyant gas (Hanna and Drivas 1987). The phenomenon was described as "slumping." If the only consideration is that the gas being dispersed has a molecular weight greater than the weight of air, then modeling the dispersion would be relatively simple. Most dense gas clouds are colder than the surrounding air and contain aerosols; therefore, the thermodynamic aspects of dispersion include consideration of (1) entrainment of warm air and condensation of water vapor, (2) evaporation and condensation of aerosol and vapor, (3) heat transfer from the ground, (4) evaporation of liquid spilled onto the ground, and (5) chemical reactions in the cloud. Chemical reactions in clouds are not usually handled in a general-purpose computer program.

Most dense gas models are based on three phases: expansion of a high-momentum jet, gravity-driven slumping, and dispersion in a neutrally buoyant cloud. The initial expansion of a flashing liquid with aerosol formation (described in Section 6.2) continues into a high-momentum jet. The jet expands and entrains air owing to the shear stresses caused by the velocity differences between the jet and the ambient air. Jet characteristics frequently are estimated by solving mass, energy, and momentum equations numerically. Entrainment coefficients based on experimental data are used to characterize interaction between the jet and the atmosphere. As a result of expansion and air entrainment, the jet velocity begins to approach the air velocity (Raj 1991, Vigeant 1991). If the jet is vertical, it will bend over when wind momentum begins to dominate jet momentum. The descending jet is like an inverted plume (Hanna and Drivas 1987).

The second phase is controlled by the density differences between the cloud and the ambient atmosphere. The weight of the cloud causes it to slump. The rate of spread of the cloud is independent of the ambient air velocity and is proportional to the square root of the difference in densities. An instantaneous release with little momentum forms a short cylinder. Air is entrained at the top and at the sides as the cylinder grows; the process again is quantified by using experimentally derived entrainment coefficients. Vertical growth initially may be negative at the center of the cylinder and positive at the edges. If the release is a plume rather than a puff, the spread in the downwind and upwind directions is restrained by material released earlier and later; therefore, the lateral spread of a plume is increased over that of a puff. The heat transfer and chemical reaction considerations listed earlier are important in this phase of dispersion because of the influence on the density-driven feature (Hanna and Drivas 1987, Vigeant 1991).

At some point, the gravitational forces are overcome by normal atmospheric turbulence, and the dispersion follows the Gaussian model. Transition from the volumetric source in the gravity-driven phase to the point source assumed by the Gaussian model is accomplished by mathematically creating a virtual point source upwind of the actual point of release. The Pasquill-Gifford or the Briggs standard deviation coefficients (σ_z, σ_y) that produce a Gaussian concentration equal to the gravity-driven concentration are determined. The distance associated with the value of σ_z and σ_y is the virtual distance. Hanna and Strimaitis (1989), CCPS (1989), and Vigeant (1991) describe the technique.

Hanna and Strimaitis (1989) compared two heavy gas models with a Gaussian model for a small, continuous chlorine release. The Gaussian model result was close to the results of the heavy gas models for a stability class D release, particularly at distances greater than 500 m. At shorter distances, the Gaussian model underpredicted the concentrations compared to dense gas models. The Gaussian model result was consistently a factor of two to four higher than the dense gas models for the example stability class F release.

Hanna et al. (1991) compared 11 gas models with 11 field experiments. Six involved continuous dense gas release, two involved instantaneous dense gas release, two involved continuous neutral-buoyancy gas release, and one involved instantaneous neutral-buoyancy gas release. No model matched the data perfectly. Some models, including the Gaussian plume model, were consistently among the better models.

The result of the dispersion analysis is the concentration of toxic or flammable material in the plume at various distances downwind, which can be expressed as concentration isopleths.

6.4 TOXIC MATERIAL EFFECTS

The toxicity of a substance, that is, the ability to cause an adverse health effect, depends on the dose-response relationship, which depends on the type of exposure and the route of entry into the body. The usual exposure mode of interest in a QRA dealing only with acute effects is the exposure to the airborne material in the plume; thus, the routes of entry are inhalation, skin absorption, and skin and/or eye contact. The ingestion route usually is consumption of contaminated food from plume deposits onto the ground. Consumption of contaminated food usually is more likely from chronic releases, not accident releases, because action is likely to be taken after the accident to prevent consumption of contaminated foods.

The potential health effects of exposure to a toxic material include irritation of eyes or mucous membranes, coughing, nausea, skin burns, neurological injury, and death. A lethal response can occur either because the concentration is very high or because mobility or judgment is impaired so that escape from a potentially less-than-lethal concentration does not occur in time to prevent lethality. These effects are immediate. Cancer and reproductive harm usually are associated with chronic exposure; however, a single exposure to a high concentration of a carcinogen may result in a delayed response (FEMA et al. undated). Decisions that result from a transportation quantitative risk analysis should target prevention of injury; however, injury is difficult to define consistently and to analyze. Most quantitative risk analyses use lethality as the criterion so that health consequence comparisons will be clear. Risk mitigation decisions based on lethality will also mitigate potential injuries.

It is difficult to be precise about human response to acute exposure to toxic materials for the following reasons (CCPS 1989). First, humans exhibit a wide range of health responses to acute exposure, as just described. Second, the severity of these effects usually varies with the time of the exposure, t, and the concentration, C. Lethal effects usually vary with the product of exposure time and the concentration raised to a power, n, that is, $C^n t$. Third, there is significant response variation among individuals, depending on health and level of activity. The elderly, the young, and those persons with respiratory or cardiovascular impairment are the most sensitive to toxic exposures. Fourth, human response usually is estimated from animal data. Fifth, the data are generally less than desirable for a single chemical; for multicomponent releases, the combined effects are more difficult to model.

Several recognized toxicological exposure limits can be used to evaluate toxic effects for transportation quantitative risk analyses, a few of which are described here (CCPS 1989, FEMA et al. undated). Emergency Response Planning Guidelines (ERPGs) for air contaminants are published by the American Industrial Hygiene Association (AIHA). The levels of ERPG are the maximum airborne concentrations for which nearly all individuals could be exposed for up to one hour *without* experiencing the following:

ERPG-1: mild, transient health effects.
ERPG-2: irreversible health effects or impairment of the ability to take protective action.
ERPG-3: life-threatening health effects.

The ERPGs do not include safety or uncertainty factors; so they may be higher than other criteria. Thirty-five ERPGs have been issued since 1984, and another 25 have been developed and are undergoing peer review before publication (Rusch 1993). These values are being developed by an industry-academia-government task force; therefore, ERPGs may become the accepted norm.

Immediately Dangerous to Life or Health (IDLH) values are published by the National Institute for Occupational Safety and Health (NIOSH). The IDLH level represents the maximum airborne concentration to which a healthy male worker can be exposed for up to 30 min without experiencing any health effects that are either escape-impairing (e.g, severe eye irritation) or irreversible. The IDLH values were published in the late 1970s for about 400 substances and have been used extensively. The Environmental Protection Agency (EPA) recommends dividing the IDLH by a factor of 10 to convert the "barely tolerable" concentration for a healthy male to a value that will protect all members of the general population.

Emergency Exposure Guidance Limits (EEGLs) and Short-term Public Emergency Guidance Levels (SPEGLs) are published by the Committee on Toxicology of the National Research Council (NRC) of the National Academy of Sciences (NAS). An EEGL concentration will neither impair the performance of tasks during emergency conditions by healthy military personnel nor result in lasting health effects. A SPEGL concentration is a concentration that will protect the public from a singie, acute exposure. The SPEGLs take into consideration the range of susceptibility of the general public. Since 1984, values for EEGLs and/or SPEGLs have been published for 43 chemicals. Exposure limits are listed in Table 6-3 for a few chemicals for illustration. Other human exposure criteria are compared in CCPS (1989) and FEMA et al. (undated).

For most chemicals the only available toxicological data are concentrations in air that cause lethality in 50% of exposed laboratory mammals (LC_{50}') or the dose that produces lethality in 50% of exposed laboratory animals (LD_{50}'), expressed as mg/kg.

IDLH, ERPG-2, EEGL, and SPEGL values are available at least for some chemicals, and they should be used if appropriate. In other cases, human

TABLE 6-3. Comparison of recognized toxicity exposure limits[1]

Chemical	IDLH[2,4]	ERPG-2[3,5]	EEGL[3,6]	Odor threshold[6]
Ammonia	500	200	100	5
Chlorine	30	3	3	0.2
Hydrogen chloride	100	20	20	0.26
Phosgene	2	0.2	0.2	0.9

Compiled from: NIOSH 1990; AIHA 1988a,b and 1990a,b; NRC 1992.

[1]All values are given in ppm. [2]Based on 30-min exposure. [3]Based on 60-min exposure. [4]NIOSH 1990. [5]AIHA 1988a,b and 1990a,b. [6]NRC 1992.

LC_{50} values determined from mammalian data should be used for lethality calculations in QRAs. Mammalian data are available from several sources including the much-used Sax and Lewis (1989). A simple procedure for determining the human LC_{50} from animal data for oral intakes, that is, LD_{50}', is:

$$LI_{50} = LD_{50}' \times M \tag{6-9}$$

$$LC_{50} = LI_{50}/(B \times t) \tag{6-10}$$

where:

LD_{50}' = lethal dose for 50% of exposed laboratory species (mg/kg)
M = mass of average-sized human, 70 kg
LI_{50} = intake producing lethality in 50% of humans exposed (mg)
LC_{50} = concentration producing lethality in 50% of humans exposed (mg/m^3)
B = human breathing rate for normal activity (3.3×10^{-4} m^3/s)
t = human exposure time of interest (s)

Toxicity data for the oral exposure route should be used with caution for QRAs involving the inhalation exposure route. The preferred approach for QRAs involving inhalation of toxic material is to use animal inhalation data. The human LI_{50} is:

$$LI_{50} = LC_{50}' \times B' \times t' \times M/M' \tag{6-11}$$

where:

LC_{50}' = lethal concentration for 50% of exposed species (mg/m^3)
B' = laboratory species breathing rate (m^3/s)
t' = laboratory exposure time to determine LC_{50}'
M' = mass of laboratory species (kg)

Then Eq. 6-10 is then used to determine LC_{50}. The LC_{50} divided by 10 is an estimate of IDLH (FEMA et al. undated).

Equations 6-9 through 6-11 are based on several important assumptions. First, intake through inhalation (the usual route under accident conditions) exhibits the same good consistency across mammalian species as does oral intake expressed in mg/kg. Second, materials that are acutely toxic at high dose rates via the oral route will be toxic via the inhalation route and vice versa. For highly toxic organic materials studied to date, less than a factor of two overall difference between oral and inhalation absorption is observed (Owen 1990). Third, the use of acute effects as a criterion for public protection will ensure protection from serious chronic effects; the exceptions to this generality should be rare. Fourth, the health effects are proportional to the product of C and t.

An alternative for some chemicals is to use probit (probability unit) values, as described next. The toxic effect for an exposure to a given concentration and duration can be obtained for about 20 chemicals using a probit equation. These equations have the distinct advantage that the exposure duration is not fixed as it is for the criteria of Table 6-3. The probit equation for lethality is of the form:

$$Pr = a + b \ln C^n t \tag{6-12}$$

where a, b, and n are material-dependent constants, C is concentration in ppm, and t is exposure time in minutes. The parameters are chosen so that Pr is a normal distribution with a mean value of 5 and a standard deviation of 1 (CCPS 1989). The probit equation value is converted to percent of population affected by using Table 6-4. For example, if the probit equation is:

$$Pr = -8.29 + 0.92 \ln C^2 t \tag{6-13}$$

then to find the percent of people affected for an exposure to a concentration of 430 ppm for 10 min, the value of Pr is calculated to be 5.0. From Table 6-4, the percent of fatalities is 50%. The equation can be used also to find the concentration associated with a specific percent of fatalities. For example,

TABLE 6-4. Transformation of percentages to probits

%	0	1	2	3	4	5	6	7	8	9
0	—	2.67	2.95	3.12	3.25	3.36	3.45	3.52	3.59	3.66
10	3.72	3.77	3.82	3.87	3.92	3.96	4.01	4.05	4.08	4.12
20	4.16	4.19	4.23	4.26	4.29	4.33	4.36	4.39	4.42	4.45
30	4.48	4.50	4.53	4.56	4.59	4.61	4.64	4.67	4.69	4.72
40	4.75	4.77	4.80	4.82	4.85	4.87	4.90	4.92	4.95	4.97
50	5.00	5.03	5.05	5.08	5.10	5.13	5.15	5.18	5.20	5.23
60	5.25	5.28	5.31	5.33	5.36	5.39	5.41	5.44	5.47	5.50
70	5.52	5.55	5.58	5.61	5.64	5.67	5.71	5.74	5.77	5.81
80	5.84	5.88	5.92	5.95	5.99	6.04	6.08	6.13	6.18	6.23
90	6.28	6.34	6.41	6.48	6.55	6.64	6.75	6.88	7.05	7.33
99	7.33	7.37	7.41	7.46	7.51	7.58	7.65	7.75	7.88	8.09

Source: Finney 1971. Reprinted with the permission of Cambridge University Press.

to find the concentration producing 50% fatalities, that is, LC_{50}, for a 10-min exposure, solve the following equation for C:

$$5.0 = -8.29 + 0.92 \ln 10C^2 \tag{6-14}$$

A value of 430 ppm results. This value can be used in the dispersion model to determine the downwind distance or the isopleth that produces 50% fatalities for those persons exposed to the plume. For a continuous release, the procedure is straightforward. For a puff release, the value of $C^n t$ in Eq. 6-12 is:

$$C^n t = \int_{t_0}^{t_f} C^n(t)\ dt \simeq \sum_{i=1}^{m} C_i^n\ \Delta t_i \tag{6-15}$$

where:

t_0 = the time at start of exposure to the puff
t_f = the time at completion of exposure to the puff

Clearly, the smaller the value of Δt_i, the more accurate the value of $C^n t$. The downwind distance and the isopleth for a specific concentration value must be found either by trial and error (selecting a location and determining the exposure) or by calculating the concentration at a number of locations and interpolating to find the location of the desired concentration. Computers are very useful for these calculations.

Withers and Lees (1985) propose Eq. 6-13 for the effects of chlorine on a population of healthy youngsters and adults who make up about 75% of the total population. Equation 6-13 is a straight line on log paper, as shown in Fig. 6-4. Withers and Lees (1985) propose that:

$$Y = -6.61 + 0.92 \ln C^2 t \tag{6-16}$$

for the vulnerable population of infants, the elderly, and those with cardiovascular or respiratory illness. Equations 6-13 and 6-16 are intended for a standard level of activity associated with an inhalation rate of 12 L/min. (For comparison, the breathing rate while walking at 2 mph is 14 L/min.) If the vulnerable population is resting in bed, the chlorine probit (Withers and Lees 1985) is:

$$Y = -7.88 + 0.92 \ln C^2 t \tag{6-17}$$

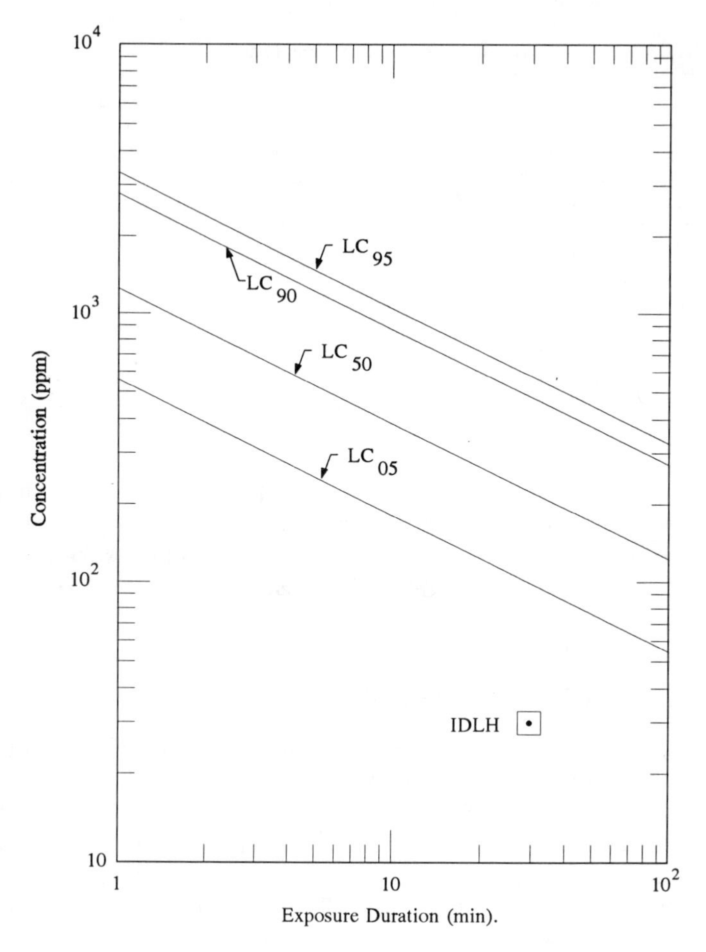

FIGURE 6-4. Lethal concentration of chlorine for different lethality probabilities as a function of exposure duration.

A comparison by Withers and Lees (1985) of the results from five probit equations for chlorine is shown in Table 6-5. A difference of almost an order of magnitude is evident at LC_{10}; this difference steadily increases with increasing LC to a factor of almost 25 at LC_{90}. This large difference in literature values for toxicity is not unique to chlorine. Hydrogen fluoride probit values that vary by a factor of 16 are reported by USEPA (1992). The large variation is due to the form of the probit equation, which makes the equation very sensitive to coefficient values and, as discussed earlier, to the difficulty in extrapolating animal data to humans.

The large uncertainty in toxic dose-response is a major reason why it is

TABLE 6-5. Comparison of chlorine fatality estimates

	Lethal Concentration for 30-min Exposure Period		
Literature source as cited by Withers and Lees (1985)	LC_{10} (ppm)	LC_{50} (ppm)	LC_{90} (ppm)
Eisenberg et al. 1975	26	34	44
Perry and Articola 1980	36	42	49
Rijnmond Public Authority 1982	237	418	738
ten Berge and van Heemst 1983	170	430	1,093
Withers and Lees 1985[a]	125	250	500

Source: Withers and Lees 1985.
[a]Eq. 6-13.

difficult if not impossible to be certain about the accuracy of absolute risk analyses (see Chapter 1). The large uncertainty in this basic parameter also brings into question the need to expend project resources to define any particular part of Eq. 2-2 in extreme detail (see also Section 2.7).

6.5 MEASURES THAT CAN REDUCE PREDICTED EXPOSURE

The calculations that are described in the first sections of this chapter often considerably overestimate the seriousness of a release when compared to the actual results when a large release occurs. The overestimation results because the models usually employ conservative assumptions and because people will take evasive actions to reduce their exposure, that is, sheltering, evacuation, and escape. Glickman and Raj (1992) illustrated the effect for releases of ammonia in two highway accidents. Without special population considerations, the calculated results were two orders of magnitude too high. The correction due to daytime/nighttime considerations was small in one case and a factor of two in the other. The correction for inside/outside populations, that is, the effect of being in a shelter, was 1.5 to 2 orders of magnitude.

For example, chlorine has a detectable odor at a concentration of 0.3 ppm. At 1 to 5 ppm, irritation of the nose, respiratory tract, and eyes begins. At 15 ppm the irritation is severe (The Chlorine Institute 1991). Chlorine gas is greenish-yellow; therefore, an exposed person generally can see the direction of the source of the gas during daylight. Under these conditions, many people can determine that evasive action is appropriate and take that action before life-threatening conditions occur. Other gases are more "sneaky"; by the time that the senses detect an exposure, a lethal exposure may have occurred. Carbon monoxide is odorless, but significant concentrations are required for lethality. Other gases, such as arsine, produce harmful effects at very low concentrations. The EEGL for arsine is 1 ppm, and it has only a mild garlic-like odor.

6.5.1 Escape

The probability of escape from a cloud containing toxic material depends on the ability of a person to detect the exposure by perceiving the odor, color, taste, or irritation properties of the material. Once one is alerted, the ratio of the detectable level to the dangerous level is important. Given that detection has occurred at sufficiently low concentrations, the proper direction for escape depends on one's knowledge of the location of the source and the wind direction (Prugh 1985). For large, instantaneous releases, the cloud might pass so quickly that detection by the senses would be inadequate to identify an appropriate escape path. Prugh gives an example of a cloud 36.6 m (120 ft) in diameter with a peak concentration of 100,000 ppm that passes in 8 s (time within the 300-ppm isopleth). The probability of escape will vary from not likely to very likely, depending on a number of variables. If a person is outdoors in the daytime, a release from a transportation accident will most likely have been preceded by noise or other potential attention-getting elements. Potentially one could avoid a lethal exposure to a cloud of toxic gas as described previously by holding one's breath and moving out of the relatively small cloud at speeds of 3 m/s (10 ft/s) for 3 to 6 s.

A simple level of analysis is to compute the isopleth for LC_{50} and to assume that the number of people inside the LC_{50} isopleth who are not fatally injured is equal to the number of people outside the LC_{50} isopleth who are fatally injured. Poblete, Lees, and Simpson (1984) show that the model is accurate when the values used in the probit reflect a high degree of homogeneity in the exposed population. Probits for fatalities from toxic gas exposure reflect less homogeneity than do fatalities or injuries from fires and explosions. If project resources are restricted, then the use of a single isopleth as described here is a very practical choice.

The possible actions of individuals affected by a toxic gas are shown in Fig. 6-5 (Purdy 1993). An analysis based on the simplest application of possible actions is to apply the isopleth model described in the previous paragraph only to those persons assumed to remain outside and a single indoor lethality calculation within the LC_{50} isopleth, as described in Section 6.5.2.

Figure 6-6 shows the model used by the British Health and Safety Commission for persons outside (Purdy 1993), which is a reasonable upper bound for the complexity of most analyses. In Zone 1 the toxic material concentration, C_1, is so high that a lethal dose is received after only a few breaths, and no escape is possible. In Zone 2, the concentration is lower, and escape indoors is possible (20% assumed). In Zone 3, escape indoors is much more likely (80% assumed). Escape by exiting the plume is considered as very unlikely

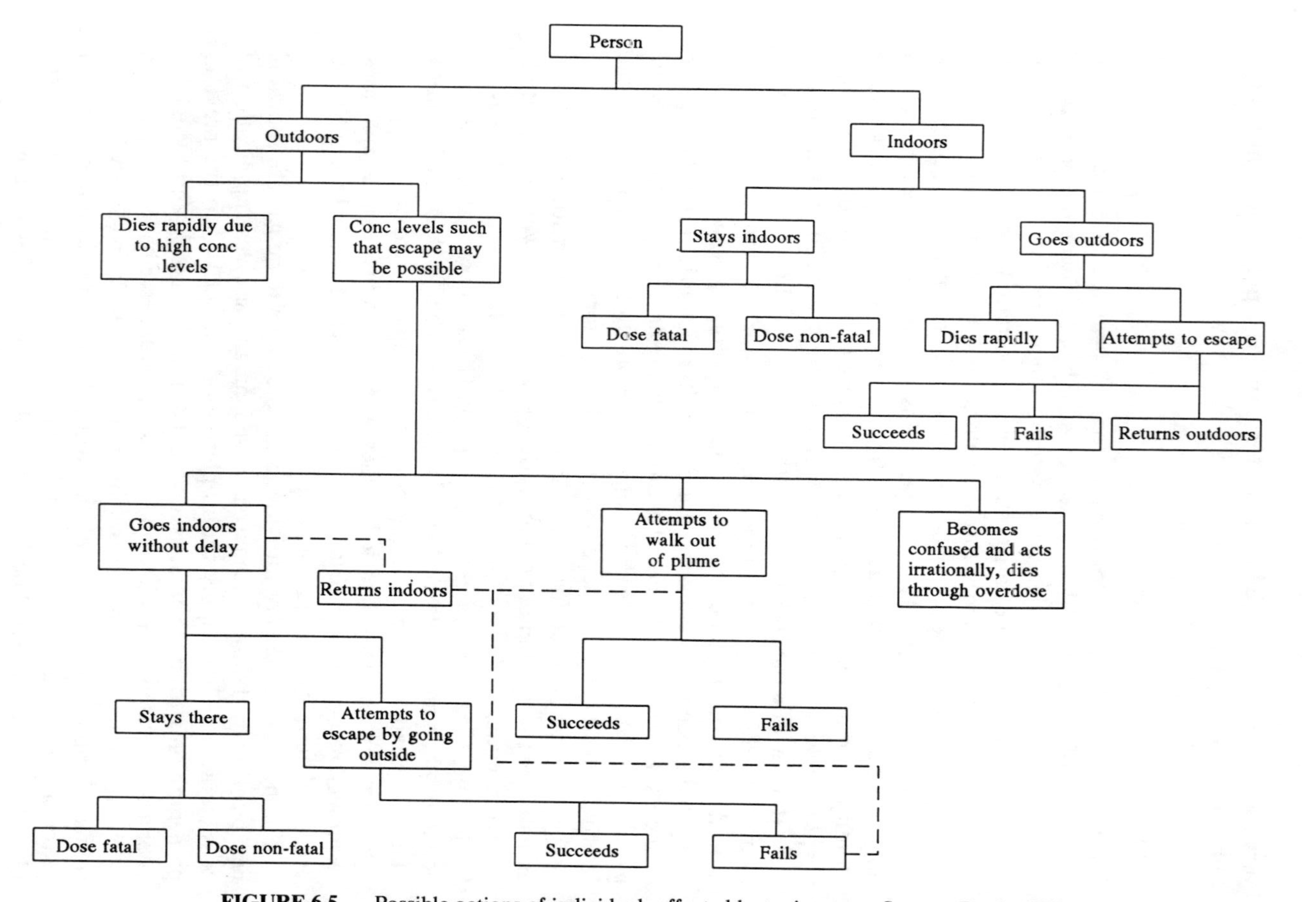

FIGURE 6-5. Possible actions of individuals affected by toxic gas. Source: Purdy 1993.

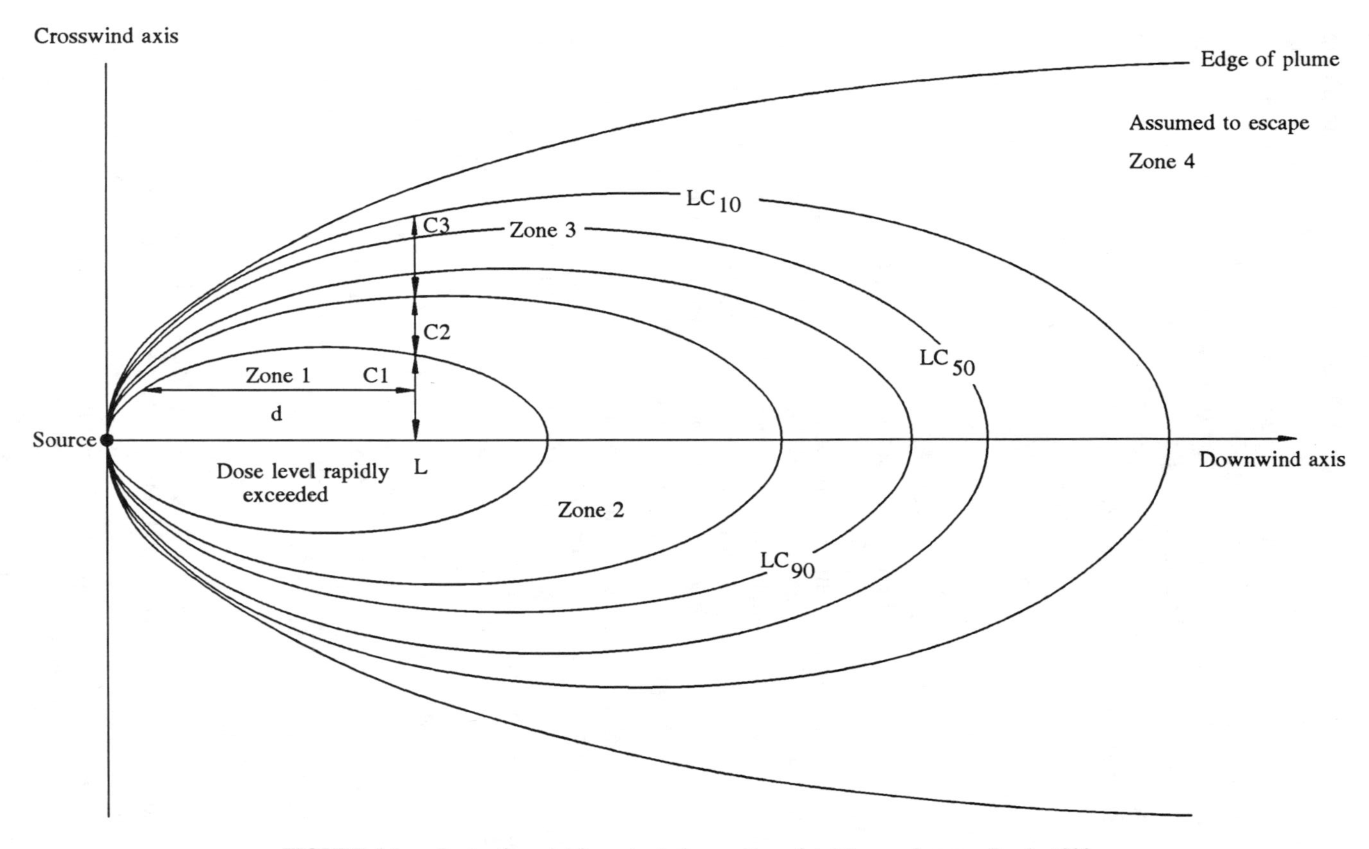

FIGURE 6-6. Basis of model for calculating outdoor fatalities. Source: Purdy 1993.

except in Zone 4, where all are assumed to escape by exiting the plume. Within Zone 3, three subzones are defined on the basis of lethal doses at the concentration-time values producing 90%, 50%, and 10% lethality, that is, LC_{90}, LC_{50}, and LC_{10}, respectively. Between LC_{50} and LC_{10}, the average lethality probability is $(0.50 + 0.10)/2 = 0.30$; similar averaging is used in the other two subzones. The number of outdoor lethalities, L_o, for a particular release scenario, r, and weather, w, assuming a uniform population density for simplicity of presentation, is:

$$L_o\,(r, w) = N \times P_o(w)\,[A_{C1} + A_{C2}\,(1 - E_{C2}) + (1 - E_{C3})\,(0.95\,A_{C3}' + 0.7\,A_{C3}'' + 0.3\,A_{C3}''')] \tag{6-18}$$

where:

N = population density (persons/km^2)
$P_o(w)$ = probability of being outdoors in weather type w
A_{C1} = area with concentration C_1 or higher (km^2)
A_{C2} = area with concentration between C_1 and C_2 (km^2)
A_{C3}' = area with concentration between C_2 and LC_{90} (km^2)
A_{C3}'' = area with concentration between LC_{90} and LC_{50} (km^2)
A_{C3}''' = area with concentration between LC_{50} and LC_{10} (km^2)
E_{C2} = probability of escape indoors in concentration zone 2
E_{C3} = probability of escape indoors in concentration zone 3

Very few data have been published on the percentage of people outside. Petts, Withers, and Lees (1987) made the following assumptions, which compared closely with wartime data from V-2 bombing: healthy youngsters and adults average 1 hr per day outside, and the vulnerable population spends 0.5 hr per day outside. The latter occurs only in the daytime. Expressed as percentages, the healthy population is outdoors 7.7% of the day and 1.4% of the night, and the vulnerable population is outdoors 4.8% of the day. On average, the population is outdoors 7.0% of the day and 1.0% of the night. Purdy et al. (1988) assumed the general population is outdoors 10% of the time in Pasquill Category D weather with a windspeed of 5 m/s (see Table 6-1) and 1% of the time in Category F weather with a windspeed of 2 m/s.

6.5.2 Sheltering

Taking shelter inside a building, closing doors and windows, shutting off ventilation systems, closing all fireplace dampers, sealing all vents, and other similar actions can significantly reduce both the peak exposure and the integrated exposure to airborne toxic materials (FEMA et al. undated). The usual indoor concentration, $C_I(t)$, model is:

$$C_I(t) = C_o\,(1 - \exp\,(-\lambda t)) \quad \text{for } t \leq t_{max} \tag{6-19}$$

and:

$$C_I(t) = C_M \exp\left[-\lambda\,(t - t_{\max})\right] \quad \text{for } t > t_{\max} \tag{6-20}$$

where:

C_o = average outdoor concentration for $t \leq t_{\max}$ (g/m^3)
C_M = indoor concentration at $t = t_{\max}$ (g/m^3)
$t_{\max}$ = time interval for cloud trailing edge to pass building (s)
λ = building ventilation rate (air changes/s)

The model is shown graphically in Fig. 6-7 and assumes that the cloud concentration is, or can be approximated by, a step increase from zero to C_o and then from C_o to zero. In Fig. 6-7, the dotted line represents the assumption that the building is evacuated because the outside concentration is less than the inside concentration. Appendix C of FEMA et al. (undated) provides equations to calculate the indoor concentration considering explicit evaluation of a ventilation system with a fresh air makeup rate, air filtration, and so on. The outside air concentration also is assumed to be constant. Extension of the model to a differential equation that can be solved by numerical integration would be the next level of complexity and would

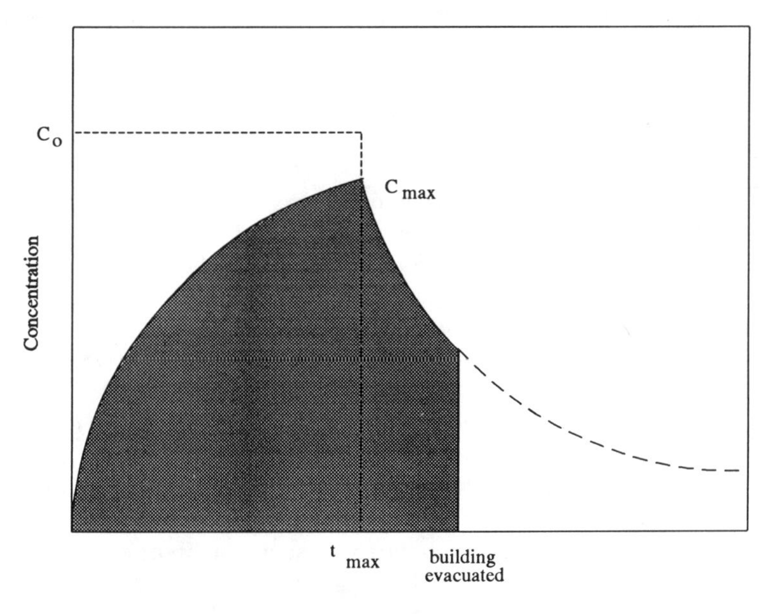

FIGURE 6-7. Simple gas infiltration model.

include a time-varying outside concentration. This level of analysis, although straightforward, is not needed in most applications.

The building ventilation rate depends on a number of factors (FEMA et al. undated), including whether doors or windows are open, the operation of mechanical ventilation system(s) such as air conditioners and exhaust fans, the air infiltration rate through cracks and small openings, and the outside wind velocity and temperature. The average ventilation rate in American homes is about 0.85 airchanges per hour (acph), and poorly constructed homes may have an average ventilation rate of 2.5 acph. An average ventilation rate for office buildings is suggested as 1 acph by FEMA et al. (undated).

The indoor lethalities, $L_i\,(r, w)$, are:

$$L_i(r, w) = [N(1 - P_o(w)) + E][0.95A_{90} + 0.7A_{50} + 0.3A_{10}] \tag{6-21}$$

where:

E = average population density of persons escaping indoors (persons/km^2)
A_{90} = area within the indoor LC_{90} isopleth (km^2)
A_{50} = area between the indoor LC_{90} and LC_{50} isopleths (km^2)
A_{10} = area between the indoor LC_{50} and LC_{10} isopleths (km^2)

Equation 6-21 was simplified by the definition of E to avoid having to keep track of the overlay of indoor and outdoor isopleths. The indoor isopleths are determined by computing the indoor concentration as a function of time (Purdy 1993).

6.5.3 Evacuation

Sheltering is the preferred action if a release has occurred. Evacuation is the preferred action if a potential future release is a threat and sufficient time is available in which to evacuate before the plume reaches the area.

Evacuations from truck and train transportation accidents and nonaccident spills averaged 59 per year (Table 2-1) from 1980 to 1984 (Sorensen 1987). A comparison of the train accident component in Table 2-1 with Federal Railroad Administration data in Table 3-23 indicates that Sorensen's data may be too low by up to a factor of two. (Sorensen found his data to be 40-50% higher than that in the HMIS; see Section 3.3.1.) In any case, on average several evacuations from truck and train accidents or incidents occur each month. For the evacuations that Sorensen investigated, it is interesting to compare the injuries due to the accident directly and those due to exposure to the released material during the evacuation, as shown in Table 6-6. No evidence of injuries or fatalities from the act of evacuation per se was found (Sorensen 1987). The increased number of injuries from incidents

TABLE 6-6. Nonoccupational exposure injuries not prevented by evacuation, by accident cause and year

Cause of evacuation	1980	1981	1982	1983	1984	Total injuries	Total incidents	Total with injuries	Percent with injuries	Average injuries per accident	Average injuries per evacuation[1]
Trail derailment	10	12	5	1	0	28	55	7	13	4	0.5
Train car spill/fire	46	0	2	2	13	63	23	5	22	12.6	2.7
Truck accident	0	0	55	3	0	58	35	4	11	14.5	1.7
Truck spill/fire	105	109	10	52	26	302	32	12	38	25.2	9.4
Chemical plant release	31	39	51	12	13	146	43	12	28	12.2	3.4
Industrial plant release	49	65	224	195	139	672	78	25	32	26.9	8.6
Pipeline	1	33	0	0	0	34	5	3	60	11.3	6.8
Ship incident	3	0	0	0	0	3	4	1	25	3	0.8
Waste site accident	0	0	0	700	0	700	7	1	14	700	100
Other	0	41	4	0	0	45	14	3	21	15	3.2
Total injuries	245	299	351	965	191	2051					
Total incidents	43	62	68	65	57	296	296				
Total with injuries	14	15	19	13	12	73		73			
Percent with injuries	33	24	28	20	21	25			25		
Average per accident	17.5	20	18.5	74	15.9	28.1				28.1	
Average per evacuation	5.7	4.8	5.2	14.8	3.4	6.9					6.9

Source: Sorensen 1987.

[1]Injuries due to exposure prior to or during evacuation and not to the evacuation per se.

(nonaccident releases) compared to accidents was attributed by Sorensen to the clear alerting signal provided by an accident, whereas an incident may be undetected for some time. For every 1000 people evacuated, 8 were injured by exposure to hazardous material prior to or during the evacuation. Some injury occurred in 25% of the evacuations.

The Mississauga evacuation of 225,000 people in a suburb of Toronto, Canada, due to a train derailment of a chlorine tank car, illustrates that evacuation can be very effective for transportation accidents. Rogers and Sorensen (1989) examined the effectiveness of emergency warning for two transportation accidents: a derailment of four cars containing hazardous material followed by a fire in Pittsburgh, Pennsylvania, on April 11, 1987, and a derailment of 21 "empty" hazardous material tank cars in Confluence, Pennsylvania, on May 6, 1987. The cumulative fraction of the population warned as a function of time is shown in Fig. 6-8. (Time zero is the start of the evacuation by officials.) At Confluence, the cumulative proportion actually evacuated followed the cumulative proportion warned curve by about 30 min, whereas, the lag time between warning and evacuation in Pittsburgh was about 30 min during the first 2 hr, but rapidly increased beyond that. As shown in Table 6-7, although 73.2% of the people at risk in Pittsburgh were

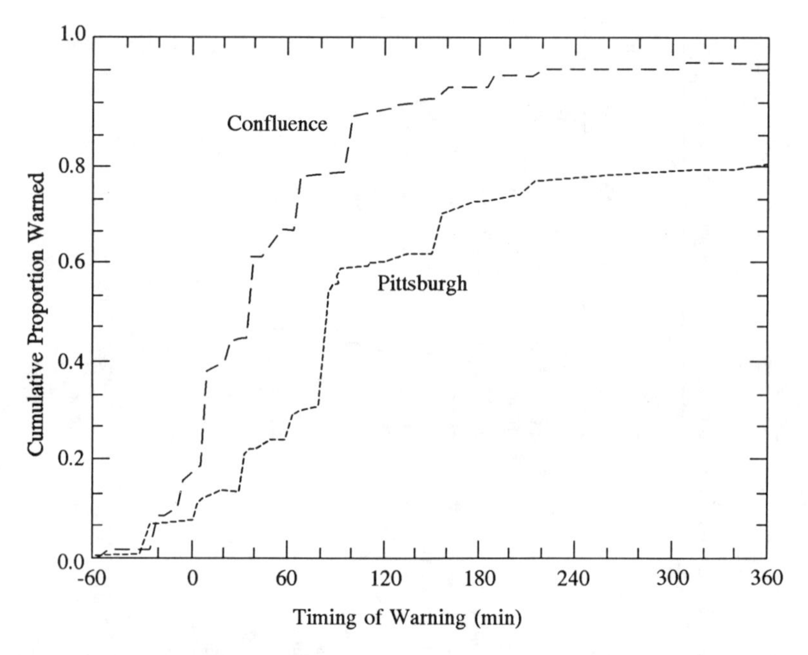

FIGURE 6-8. Cumulative population warned as a function of time following two chemical accidents. Source: Sorensen 1990.

TABLE 6-7. Public response to emergency warning summary

	Pittsburgh	Confluence	Mississauga[1]	Mt. Vernon	Denver
Population at risk	16,000-22,000	986	3500	3750	4900
Percent warned	73.2	90.5	99	82	96
How warned	Route/door	Route/door	Route/door	Route/door	Route/door
Percent warned in first hour	23.4	68.4	NA	NA	NA
Total warning time (hours)	NA[2]	NA	2	2.5	2.5
Percent evacuated	40	85	98	67	82
Mean response time (minutes)	26.5	24.2	NA	NA	NA

Source: Rogers and Sorensen 1989.
[1]First area to be evacuated. [2]NA = not available.

warned, only 40% ever evacuated. At Confluence, 90.5% were warned, and 85% evacuated. Rogers and Sorensen (1989) found that the response is a function of the warning message and the source of the information. Sorensen and Mileti (1989) report that given a minimum of 3 to 4 hr warning lead time, at least 90% of the population can be warned without the use of a highly specialized warning system. Prugh (1985) estimates evacuation effectiveness (expressed as failure to evacuate) as a function of warning time, the evacuation area, and the population density, as shown in Fig. 6-9. The uncertainty is estimated by Prugh as "within a factor of two or so." Prugh's curve does not show the effect of short warning times corresponding to people very near the

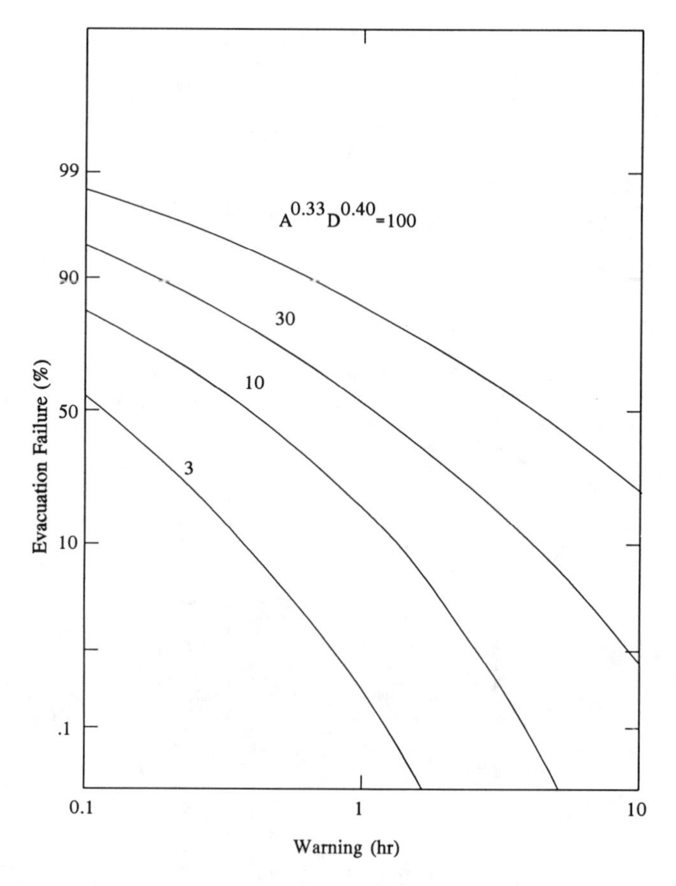

FIGURE 6-9. Evacuation failure as function of warning time, the area A to be evacuated (square miles), and population density D (people per square mile). Source: Prugh, R. W. April 1985. Mitigation of vapor cloud hazards. *Plant/Operations Progress* 4(2):95–103. Used by permission of the American Institute of Chemical Engineers. ©AIChE. All rights reserved.

hazard source. People very near the accident will be warned and start evacuating before officials initiate a formal warning and evacuation, as shown in Fig. 6-8.

6.5.4 Summary

A lack of consideration of evasive action usually will lead to an overestimation of the number of fatalities. Many risk assessments do not model evasive actions because of the large element of judgment required. A reasonable estimate of the net fatalities, given a fatal exposure and considering the chemical(s) involved, is probably the most cost-effective approach until better data become available. Analyses by Glickman and Raj (1992) indicate that for ammonia between 1% and 2% of the calculated fatal exposures actually are fatalities. A simple assumption that all persons exposed to LC_{50} and higher are fatalities is balanced by those exposed to less than LC_{50} who would, in fact, become fatalities.

6.6 POPULATION EXPOSED

Results of the 1990 U.S. census are available from The Bureau of the Census. Maps for each county show the lowest unit for which data are available: the block. Blocks usually are bounded by streets or other visible features. In large cities, the block may be an actual city block, but in less populated places, areas the size of several city blocks are grouped into a single block. In some cases, the blocks are best defined in maps of special designations such as a metropolitan statistical area (U.S. Bureau of the Census 1991).

The census bureau data are for the places where people live. During normal working hours, the population in residential areas will be smaller, and the population in business areas will be greater than the population for the same areas at other times. A general expression for a residential area is:

$$\overline{N} = N\left[\frac{5}{7}\left(\frac{10}{24}[1 - f_e^{\text{out}} + f_e^{\text{in}}] + \frac{14}{24}\right) + \frac{2}{7}\right] \tag{6-22}$$

where:

$\overline{N}$ = average population (persons/km^2)
N = census bureau block data (persons/km^2)
f_e^{out} = fraction of population working outside the block
f_e^{in} = population living outside the block but working inside the block, expressed as a fraction of those living in the block

The first term accounts for a 10-hr absence, 5 days a week; the second term accounts for the other 14 hr of each weekday; and the third term accounts for the weekend. In a series of figures Glickman (1986) describes population departures, arrivals, and averages for commercial, residential, and mixed areas by time of day in 1-hr increments.

A three-zone population calculation model has been proposed by Petts et al. (1987). For the first 400 m downwind, the population estimation should be as accurate as project resources permit: census data checked by visual inspection and occasional direct inquiry at houses or buildings should be sufficiently accurate. The second zone extends from 400 m to 1,000 m. In this zone, maps such as the 1:24,000 scale U.S. Geological Survey maps can be used to estimate high, medium, low, and zero population areas. Estimated average values for the population categories based on the local characteristics should be used. At distances greater than 1,000 m, national averages for urban/suburban, rural, and uninhabited areas can be used. The areas with these characteristics also can be estimated from maps.

A major highway accident usually causes a substantial accumulation of vehicles in the lane(s) behind the wreck, that is, persons traveling in the same direction as the wrecked vehicle(s). Vehicles traveling in the opposite direction almost always will slow down and rubberneck. If hazardous material is released, it is likely that traffic in both directions will be stopped. At an intersection, the effect is compounded. In a recent study by the Health and Safety Commission's Advisory Committee on Dangerous Substances (HSC 1991) covering transportation of hazardous materials in Great Britain, the major risk for LPG, chlorine, and ammonia was to other motorists. Of the materials examined, only explosives and gasoline did not share this feature. Using average vehicle lengths and occupancies, an average motorist density of 0.05 persons/m^2 was calculated by the HSC.

In the United States, hourly traffic counts usually are available from the Departments of Transportation of the respective states. An average vehicle occupancy is estimated at 1.16 by Glickman and Raj (1992).

6.7 FLAMMABLE MATERIAL ANALYSIS

If the material being transported is flammable, then the potential exists for a fire or an explosion. The number of possible outcomes are many, as illustrated in Fig. 6-10 for a flammable gas release and Fig. 6-11 for a flammable liquid release (CCPS 1989). The outcomes are described briefly in this section. Presentation of the detailed modeling requirements of the consequences of these fires and explosions is beyond the scope of this book, but the transportation risk analysis decision maker can use this summary to better understand options and to choose analysts who have the desired

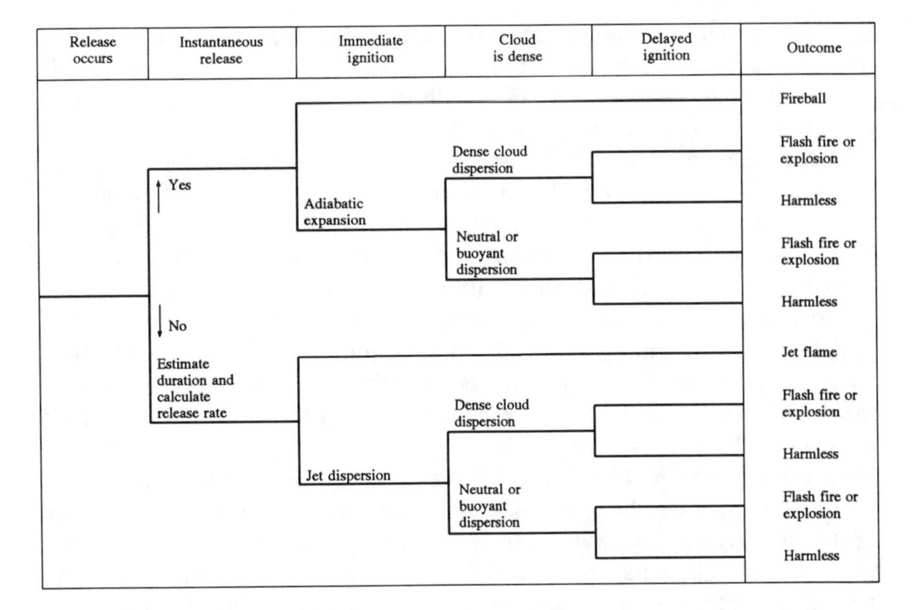

FIGURE 6-10. Flammable gas event tree. Adapted from: CCPS 1989.

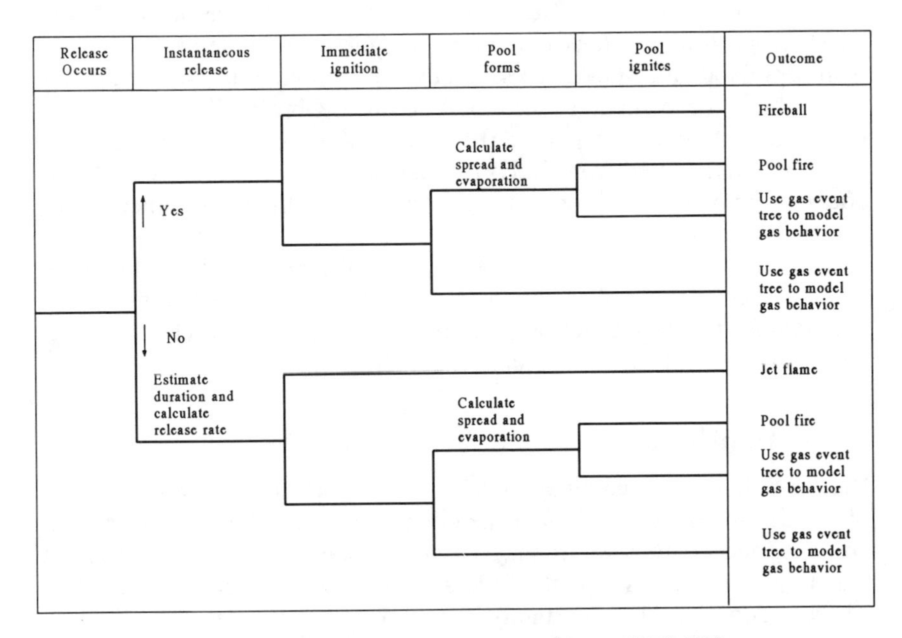

FIGURE 6-11. Flammable liquid event tree. Adapted from: CCPS 1989.

modeling capability. Several good sources of detailed analytical modeling information are available, including CCPS (1989), the "Yellow Book" (TNO 1992a), and the "Green Book" (TNO 1992b). The information presented in this section is summarized from CCPS.

A sudden release of a large quantity of flammable material may be ignited before it is diluted below its lower flammability limit (LFL). The cloud may start out above the upper flammability limit (UFL); that is, the air/fuel mixture is too rich. Alternatively, a small, continuous release may build up to the LFL and then ignite. A flammable vapor cloud may be ignited from a number of sources including cigarettes, vehicles, or electrical systems. The cloud normally is ignited at the edge. Early ignition while the cloud is small usually will have a lesser consequence than that of a later ignition when the maximum size in the UFL/LFL range occurs. An unconfined vapor cloud explosion (UCVE) results if the flame front propagates through the cloud at high speeds. The precise nature of the UCVE is not well understood; a detonation or, more likely, a deflagration may occur. The main consequence of the UCVE is the shockwave, and consequences frequently are calculated from the equivalent mass of TNT. If the flame front propagates slowly, a flash fire occurs. The main consequence of a flash fire is thermal radiation.

If a sudden release of a large quantity of flammable material is ignited immediately, a fireball results. A BLEVE may be the source of the sudden release. Most documented BLEVEs are caused by a fire impinging on the unwetted portion of a tank containing a liquid at a temperature above its boiling point at atmospheric pressure; thus, if the material inside the tank is flammable, a fireball will result when the release occurs. The BLEVE will propel tank fragments up to 500 m away. Many deaths and secondary releases have resulted from BLEVE tank fragments. BLEVE overpressure effects usually are of secondary importance; the main consequence of a fireball is thermal radiation.

The effects of pool fires and jet fires usually are localized. The main consequence is thermal radiation. A very important consideration is the potential for the pool or jet fire to impinge on adjacent sources of toxic or flammable material and cause additional release(s).

6.8 SUMMARY

Risk is defined as frequency and consequence; that is, consequence is one-half of the risk expression. Depending on the use of the risk analysis, the consequence analysis effort may be substantially greater than, approximately equal to, or substantially less than the frequency analysis effort. There is considerable uncertainty in quantitative estimates of both consequences and frequencies. In consequence analysis, factors such as the degree of analytical detail needed for the decision, the project resources, and the inherent uncertainty because of poor toxicity data may suggest a less detailed frequency analysis approach. Similarly, the approximations made in the frequency analysis owing to project constraints may suggest that a less detailed consequence analysis would be appropriate.

References

Note: The reports of U.S. government agencies, their laboratories, and contractors cited here are available from the National Technical Information Service, Springfield, Virginia 22161, USA.

AIHA (American Industrial Hygiene Association). 1988a. *Emergency Response Guidelines, Chlorine.* Akron, Ohio.

AIHA (American Industrial Hygiene Association). 1988b. *Emergency Response Guidelines, Ammonia.* Akron, Ohio.

AIHA (American Industrial Hygiene Association). 1990a. *Emergency Response Guidelines, Phosgene.* Akron, Ohio.

AIHA (American Industrial Hygiene Association). 1990b. *Emergency Response Guidelines, Hydrogen Chloride.* Akron, Ohio.

Briggs, G. A. 1973. *Diffusion Estimation for Small Emissions.* File No. 79. Atmospheric Turbulence and Diffusion Laboratory. Oak Ridge, Tennessee.

CCPS (Center for Chemical Process Safety). 1989. *Guidelines for Chemical Process Quantitative Risk Analysis.* New York: American Institute of Chemical Engineers.

Clough, P. N., D. R. Grist, and C. J. Wheatly. November 1987. The mixing of anhydrous hydrogen fluoride with moist air. In proceedings of *International Conference on Vapor Cloud Modeling,* Cambridge, Maryland, pp. 39-55. New York: American Institute of Chemical Engineers.

Eisenberg, N. A., C. J. Lynch, and R. J. Breeding. June 1975. *Vulnerability Model, a Simulation System for Assessing Damage Resulting from Marine Spills.* AD-A015245. U.S. Coast Guard.

Fauske, H. K., and M. Epstein. April 1988. Source term considerations in connection with chemical accidents and vapor cloud modeling. *Journal of Loss Prevention Process Industry* 1(2):75-83.

FEMA (Federal Emergency Management Agency), U.S. Department of Transportation, and U.S. Environmental Protection Agency. Undated. *Handbook of Chemical Hazard Analysis Procedures.*

Finney, D. J. 1971. *Probit Analysis.* Third Edition. Cambridge, England: Cambridge University Press.

Gifford, F. A., Jr. July 1968. An outline of theories of diffusion in the lower layer of the atmosphere. In *Meteorology and Atomic Energy,* ed. D. H. Slade. TID-24190. U.S. Atomic Energy Commission.

Gifford, F. A., Jr. 1976. Turbulent diffusion-typing schemes: a review. *Nuclear Safety* 17(1):68-86.

Glickman, T. S. 1986. A methodology for estimating time-of-day variations in the size of a population exposed to risk. *Risk Analysis* 6(3):317-324.

Glickman, T. A., and P. K. Raj. 1992. A comparison of theoretical and actual consequences in two fatal ammonia incidents. Paper read at International Consensus Conference on the Risks of Transporting Dangerous Goods, April 6-8, Toronto, Canada.

Hanna, S. R., and P. J. Drivas. 1987. *Guidelines for Use of Vapor Cloud Dispersion Models.* New York: Center for Chemical Process Safety of the American Institute of Chemical Engineers.

Hanna, S. R., and D. Strimaitis. 1989. *Workbook of Test Cases for Vapor Cloud Source Dispersion Models.* New York: American Institute of Chemical Engineers.

Hanna, S. R., G. A. Briggs, and R. P. Hosker, Jr. 1982. *Handbook on Atmospheric Diffusion.* DOE/TIC-11223. U.S. Department of Energy.

Hanna, S. R., D. G. Strimaitas, and J. C. Chang. May 1991. Uncertainties in Hazardous Gas Model Predictions. In proceedings of the *International Conference and Workshop on Modeling and Mitigating the Consequences of Accidental Releases of Hazardous Materials,* New Orleans, Louisiana, pp. 348-368. New York: American Institute of Chemical Engineers.

Hosker, R. P., Jr. 1982. *Methods for Estimating Wake Flow and Effluent Dispersion Near Simple Block-like Buildings.* NUREG/CR-2521. U.S. Nuclear Regulatory Commission.

HSC (Health and Safety Commission, Advisory Committee on Dangerous Substances). 1991. *Major Hazard Aspects of the Transport of Dangerous Substances.* London: Her Majesty's Stationery Office.

Johnson, D. W. May 1991. Prediction of aerosol formulation from the release of pressurized, superheated liquids to the atmosphere. In proceedings of the *International Conference and Workshop on Modeling and Mitigating the Consequences of Accidental Releases of Hazardous Materials,* New Orleans, Louisiana, pp. 1-34. New York: American Institute of Chemical Engineers.

Leslie, I. R. M., and A. M. Birk. 1991. State of the art review of pressure liquified gas container failure modes and associated projectile hazards. *Journal of Hazardous Materials* 28(3):329-365.

McDevitt, C. A., et al. 1990. Initiation step of boiling liquid expanding explosions. *Journal of Hazardous Materials* 25(1):169-180.

NIOSH (National Institute for Occupational Safety and Health). June 1990. *NIOSH Pocket Guide to Chemical Hazards.* Publication 90-117. U.S. Department of Health and Human Services.

NRC (National Research Council). 1992. *Compilation of Current EEGLs and CEGLs. Committee on Toxicology.*

Owen, B. A. 1990. Literature-derived absorption coefficients for 39 chemicals via oral and inhalation routes of exposure. *Regulatory Toxicology and Pharmacology* 11:237-252.

Pasquill, F. 1961. The estimation of the dispersion of windborne material. *Meteorological Magazine* 90:33-49.

Perry, R. H., and D. W. Green, eds. 1989. *Perry's Chemical Engineers' Handbook.* Sixth Edition. New York: McGraw-Hill.

Perry, W. W., and W. P. Articola. 1980. *Study to Modify the Vulnerability Model of the Risk Management System.* AD-A084214. U.S. Coast Guard.

Petts, J. I., R. M. J. Withers, and F. P. Lees. 1987. The assessment of major hazards: the density and other characteristics of the exposed population around a hazard source. *Journal of Hazardous Materials* 14(3):337-363.

Poblete, B. R., F. P. Lees, and G.B. Simpson. 1984. The assessment of major hazards: estimation of injury and damage around a hazard source using an impact model based on inverse square law and probit relations. *Journal of Hazardous Materials* 9(3):355-371.

Prugh, R. W. April 1985. Mitigation of vapor cloud hazards. *Plant/Operations Progress* 4(2):95-103.

Prugh, R. W. February 1991. Quantify BLEVE hazards. *Chemical Engineering Progress* 87(2):66-72.

Purdy, G. 1993. Risk analysis of the transportation of dangerous goods by road and rail. *Journal of Hazardous Materials* 33(2):229-259.

Purdy, G., H. S. Campbell, G. C. Grint, and L. M. Smith. 1988. An analysis of the risks arising from the transport of liquefied gases in Great Britain. *Journal of Hazardous Materials* 20(3):335-355.

Raj, P. K. May 1991. Chemical release/spill source models—a review. In proceedings of the *International Conference and Workshop on Modeling and Mitigating the Consequences of Accidental Releases of Hazardous Materials,* New Orleans, Louisiana, pp. 87-102. New York: American Institute of Chemical Engineers.

Rijnmond Public Authority. 1982. *Risk Analysis of Six Potentially Hazardous Industrial Objects in the Rijnmond Area, a Pilot Study.* Dordrecht, Holland: D. Reidel Publishing Co.

Rogers, G. O., and J. H. Sorensen. 1989. Warning and response in two hazardous materials transportation accidents in the U.S. *Journal of Hazardous Materials* 22(1):57-74.

Rusch, G. M. 1993. The history and development of emergency response planning guidelines. *Journal of Hazardous Materials* 33(2):193-202.

Sakiadis, B. C. 1984. Fluid and particle mechanics. In *Perry's Chemical Engineers' Handbook,* Sixth Edition, ed. R. H. Perry, D. W. Green, and J. O. Maloney. New York: McGraw Hill.

Sax, N. I., and R. J. Lewis, Sr. 1989. *Dangerous Properties of Industrial Materials.* Seventh Edition. New York: Van Nostrand Reinhold.

Sorensen, J. H. 1987. Evacuations due to off-site releases from chemical accidents: experience from 1980 to 1984. *Journal of Hazardous Materials* 14(2):247-257.

Sorensen, J. H., and D. S. Mileti. September 1989. Warning systems for nuclear power plant emergencies. *Nuclear Safety* 30(3):358-370.

ten Berge, W. F., and M. V. van Heemst. 1983. Validity and accuracy of a commonly used toxicity model in risk analysis. In *Proceedings of the Fourth International Symposium on Loss Prevention in the Process Industries,* Vol. 1, p. 11. Rugby, United Kingdom: Institute of Chemical Engineering.

The Chlorine Institute. July 1991. *Chlorine and Your Health. . . .* Second Edition. Washington, D.C.: The Chlorine Institute.

TNO (The Netherlands Organization of Applied Scientific Research). 1992a. *Methods for the Calculation of the Physical Effects Resulting from Releases of Hazardous Materials (Liquids and Gases).* Second Edition. CPR 14E. The Hague, Netherlands: The Director-General of Labour.

TNO (The Netherlands Organization of Applied Scientific Research). 1992b. *Meth-*

ods for Determination of Possible Damage to People and Objects Resulting from Releases of Hazardous Materials. 16E. The Hague, Netherlands: The Director-General of Labour.

Turner, D. B. 1964. A diffusion model for an urban area. *Journal of Applied Meteorology* 3(1):83-91.

Turner, D. B. 1970. *Workbook of Atmospheric Dispersion Estimates.* PB-191482. U.S. Dept. of Health, Education, and Welfare.

U.S. Bureau of the Census. March 1991. Census geography—concepts and products. In *Factfinder for the Nation.* CFF No. 8 (Rev.), pp. 1-8.

USEPA (U. S. Environmental Protection Agency). Office of Solid Waste and Emergency Response. May 1992. *Report to Congress, Section 301(17)(6) Clean Air Amendments of 1990.* Unnumbered draft report. Office of Solid Waste and Emergency Response.

Vigeant, S. A., 1991. Chemical plume dispersion analysis. In *Risk Assessment and Risk Management for the Chemical Process Industry,* ed. H. R. Greenberg and J. J. Cramer. pp. 167-195. New York: Van Nostrand Reinhold.

Withers, R. M. J., and F. P. Lees. 1985. The assessment of major hazards: the lethal toxicity of chlorine, Part 2, Model of toxicity to man. *Journal of Hazardous Materials* 12(3):283-302.

7

Risk Determination and Presentation

The preceding chapters discussed how accident scenarios are developed, how the accident release portions of the scenario frequencies are determined, and how the scenario consequences are calculated. Section 7.1 explains how to complete the overall frequency calculation by combining the accident release frequency with the conditional probabilities arising from the consequence calculations. The result of the risk analysis is a large number of sets of three data elements, or triplets: the overall release and exposure scenario, the overall scenario frequency, and the scenario consequence. These triplets are very useful to show which scenarios contribute the most to the risk and therefore where to apply resources most effectively to reduce risk. The sheer number of triplets generated makes direct presentation of this information impractical for most situations. The two most common risk measures for presentation are individual risk and societal risk; these risk measures and their presentation formats are described in Section 7.2. A summary discussion of the characteristics of each presentation format is contained in Section 7.3.

7.1 RISK DETERMINATION

The quantitative risk analysis procedure develops accident scenarios leading to a release of hazardous materials (Chapter 4). The frequency of occurrence of each of these release scenarios is estimated by using data from Chapter 3 and failure threshold evaluations as described in Chapter 5. At this stage each scenario is developed to the point that a release occurred, but the size of the release and other consequence parameters have not yet been associated with the scenario. Thus, each release scenario can have several outcomes, each with different probabilities. For example, a hole produced by accident forces might be large or small and might be in the vapor space or the liquid space of the container. Another example of variable outcomes that affect overall scenario frequency is the potential actions of individuals in the downwind direction of a toxic plume (see Fig. 6-5). The weather also will affect dispersion parameters important for toxic material releases, and two or three sets of weather parameters can be considered in the consequence calculation along with their respective probabilities, which will affect the overall frequency value.

The wind direction will determine which population is exposed to a downwind plume, and a wind rose is used to plot the probability that the wind will come from a particular direction and at a particular speed. A typical wind rose is shown in Fig. 7-1, but in most transportation accident scenarios no data are available for the variation in wind direction. Lacking data, a uniform wind rose is assumed. If eight directions are used, the probability of each of the eight directions is 0.125.

All of these considerations, and possibly others for specific situations,

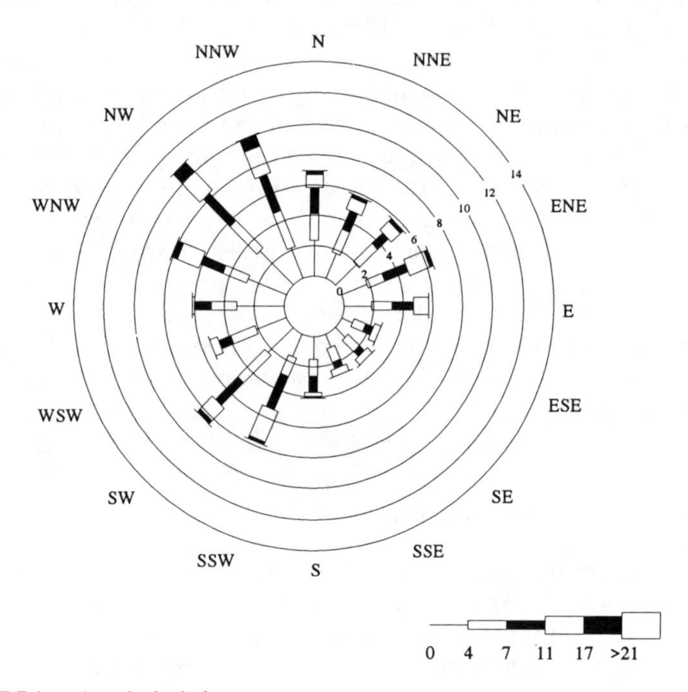

FIGURE 7-1. A typical wind rose.

affect the overall scenario frequency. A typical expression for the overall frequency of a specific outcome affecting a specific downwind location is:

$$F_i\,(x,y) = (F_{i,r})(P_{i,o})(P_{i,p})(P_{i,e})(P_{i,w})(P_{i,h}) \tag{7-1}$$

where:

$F_i\,(x,y)$ = the frequency of a health effect from outcome scenario i at location x,y

$F_{i,r}$ = the frequency of release scenario i producing a release from the transport container, evaluated using data and methods from Chapters 3, 4, and 5

$P_{i,o}$ = the probability of an outcome from release scenario i that will lead to the health effect of interest, evaluated from Chapter 6

$P_{i,p}$ = the probability that a certain population will be present, including the effects of parameters such as day/night, indoor/outdoor, healthy/vulnerable, etc., evaluated from Chapter 6

$P_{i,e}$ = the probability that evasive action e will be taken after the occurrence of release scenario i, evaluated from Chapter 6

$P_{i,w}$ = the probability of weather type w, used in dispersion equations to produce a concentration, $C_{(x,y)}$, at the location x,y (including the effect of sheltering, if any), evaluated from Chapter 6

$P_{i,h}$ = the probability of health effect h from concentration $C_{(x,y)}$, evaluated from Chapter 6

For transportation quantitative risk analysis, the route is divided into increments in which the accident rate, population, and other parameters contributing to the risk calculation can be approximated adequately by a uniform distribution. If there are ten release scenarios, three outcome cases of interest for each scenario, two evasive action choices for each release, three weather types affecting the outcome concentration of flammable or toxic material, and three concentration isopleths of interest, then the total number of cases is $10 \times 3 \times 2 \times 3 \times 3 = 540$ at each accident location along the transport route. Each case consists of a triplet: the scenario describing the release and exposure scenario, the frequency, and the consequence. These triplets can be very useful to the technical analyst to show which scenarios are contributing the most to the risk and thus are prime candidates for reducing the risk by their mitigation. Presentation of this large amount of information to the lay decision maker, however, generally is impractical.

7.2 RISK MEASURE AND PRESENTATION

The two types of risk measures are individual risk and societal risk. Individual risk reflects the frequency of a specific health effect at a specific geographical location. Individual risk calculations are used to answer the question: How does the risk to an individual vary with location? Two presentation techniques generally are used for individual risk: risk contours that are isopleths of constant individual risk shown in a map format and risk profiles that show individual risk as a function of distance. In both cases, the expected frequency of the specified health effect is shown whether or not anyone is present at the location. A risk contour is shown in Fig. 7-2 and a risk profile in Fig. 7-3.

Societal risk reflects the frequency of health effects (usually fatalities) in a specific population occurring as the result of exposure to a specified hazardous material. The most common societal risk presentation technique is an *F-N* curve, which is a graph of number of fatalities (*N*) on the abscissa and

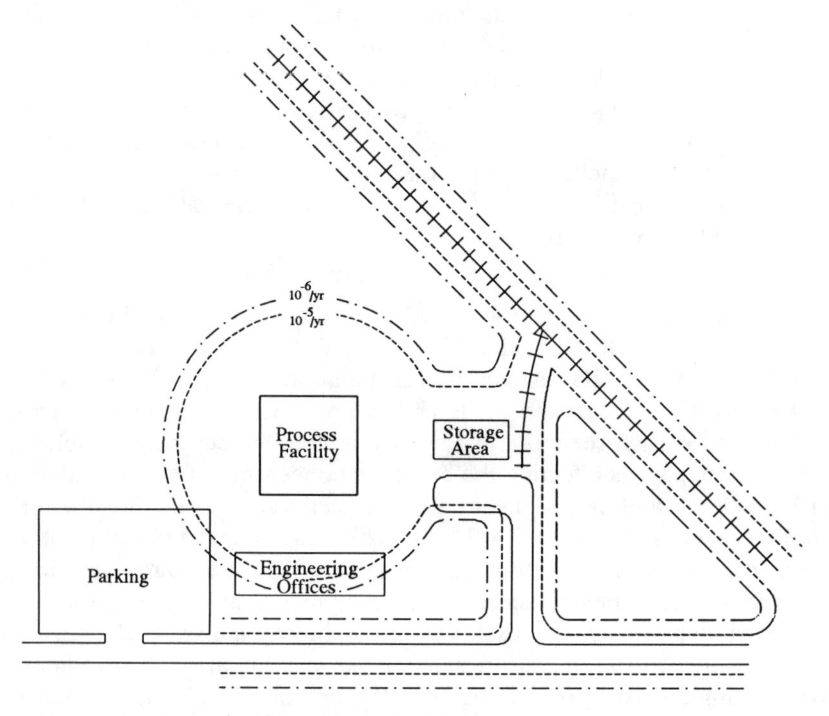

FIGURE 7-2. Example individual risk contours showing frequency of a fatality.

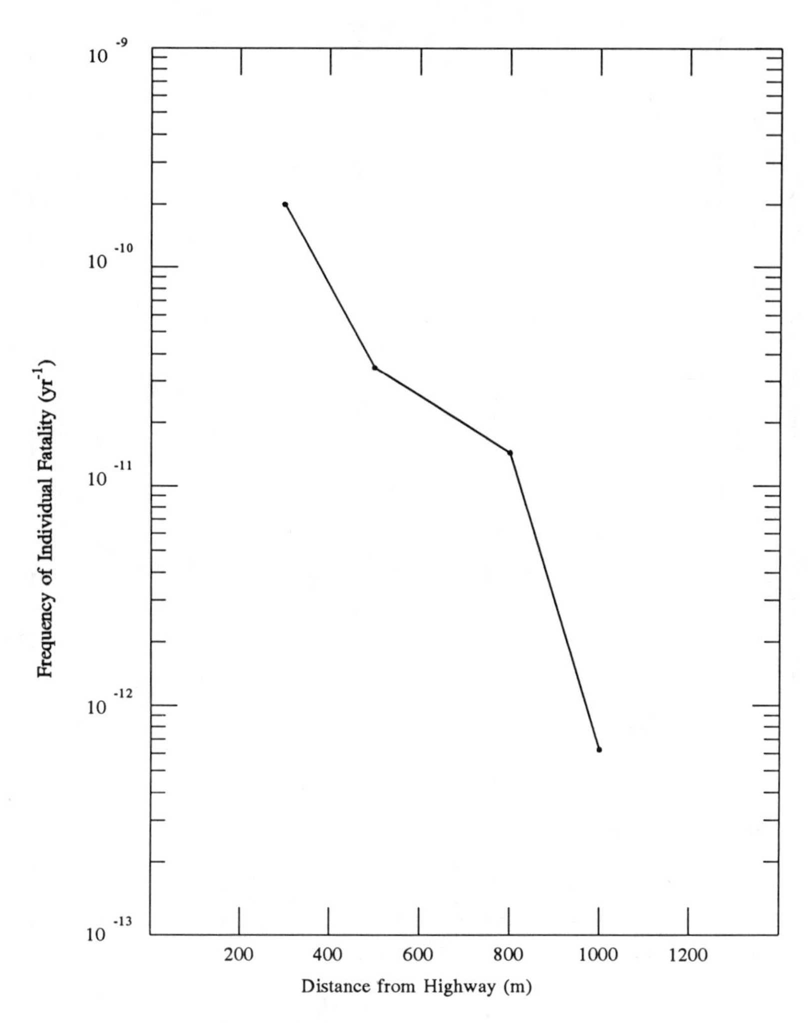

FIGURE 7-3. Example of individual risk profile.

frequency (F) of N or more fatalities on the ordinate. An example F-N curve is shown in Fig. 7-4.

7.2.1 Individual Risk Calculation

The individual risk at location x,y is the sum of the risks from all scenarios producing a common health effect, usually fatalities. (For the remainder of this chapter the health effect of interest will be fatalities. See the discussion

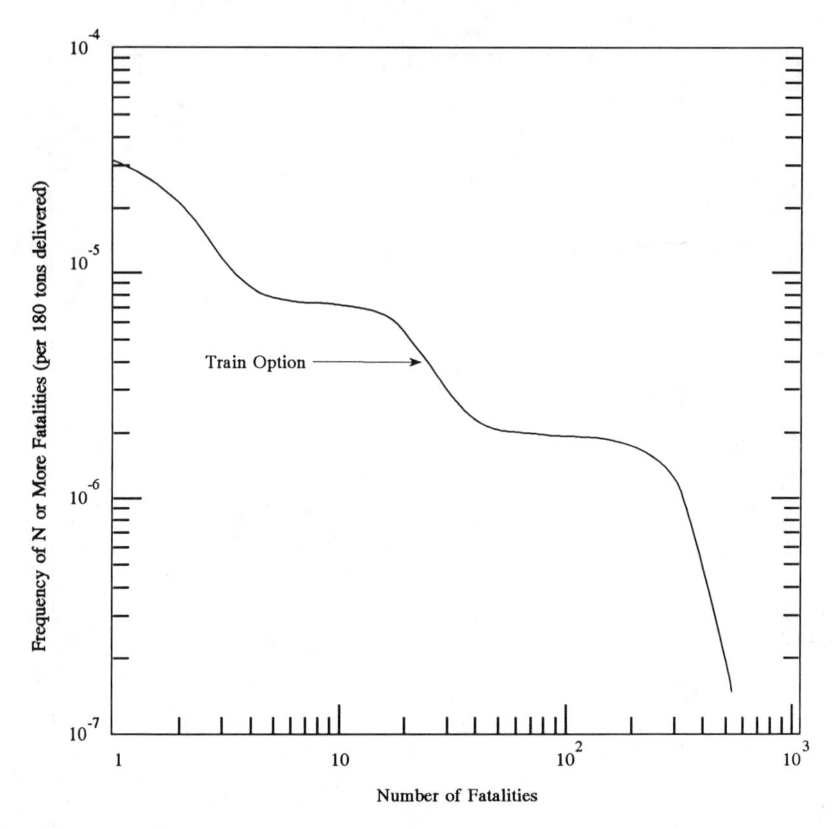

FIGURE 7-4. Example of societal risk *F-N* curve.

of toxic material health effects in Section 6.4.) For the consequence of an individual fatality, Eq. 2-1 can be written as:

$$R_I(x,y) = \sum_{i=1}^{i_{max}} F_i(x,y) \tag{7-2}$$

where $F_i\,(x,y)$ is evaluated from Eq. 7-1 and may involve additional summations for the r, o, e, w, and h indices, and $R_I\,(x,y)$ is the frequency of a fatality at location x,y.

The individual risk at a geographic location x,y is the sum of the contributions from all scenarios as given by Eq. 7-2 at all accident locations along the route. In practice, the route is divided into segments, and the segment lengths typically are longer than the lethal distance arising from the worst scenario; therefore, multiplying the route segment length by the accident rate will overestimate the risk frequency contribution. The proper approach is to compute the risk from all scenarios occurring at a single point and then integrate along the route segment. Mathematically, the risk along a route segment is:

$$R_I(x,y) = \int_{x_1}^{x_2} \sum_{i=1}^{i_{\max}} F_i(x,y)\, dx \tag{7-3}$$

where the route segment starts at x_1 and ends at x_2. The evaluation of the integral is not straightforward because the angle of the wind direction from the postulated accident site to the desired x,y location and the downwind distance between the postulated accident location and the point of exposure both vary with x. An assumption that there is no azimuthal dependence to the wind direction would still leave the problem of the downwind distance variation. If an analytical expression were available for the downwind concentration, then Eq. 7-3 could be integrated analytically. As this is not the case, Eq. 7-3 must be approximated in some fashion.

The effect area from a single lethality isopleth for a single scenario at multiple locations along a route segment is illustrated in Fig. 7-5. The point x,y is within isopleths originating at points 1, 2, and 3 but not at point 4. The isopleths are of identical shape; this implies that there is no azimuthal dependence of the isopleth calculation. For ease of presentation, this discussion will be based on that assumption. One solution to Eq. 7-3 is to observe that at locations farther than a distance of d_1 from point 1, the single scenario being evaluated makes no contribution to risk along the centerline of the plume originating at point 1. A first approximation to Eq. 7-3 is to compute the accident frequency that can affect the point x,y as twice the lethality distance (d_1) of the specific scenario times the accident rate. For points near the route, the approximation is good; but as y approaches d_1 (in Fig. 7-5), the approximation accuracy decreases rapidly. A better approximation with little additional computation cost is to use the chord distance at x,y rather than twice the radius (Saccomanno and Shortreed 1991). The chord length at a distance y from the center of a circle of radius r is:

$$\text{chord length} = 2\sqrt{r^2 - y^2} = 2r\sqrt{1 - f^2} \tag{7-4}$$

where $y = f$ times r, $0 \leq f \leq 1$. The chord length rather than the segment length should be used with the accident rate to determine $F_{i,r}$ in Eq. 7-1. Alternatively, the segment length times the accident rate can be used as the base accident frequency, and the calculation of the individual risk can proceed with a correction factor to $F_{i,r}$ in Eq. 7-1 of the chord length divided by the segment length. The latter approach is used here. As the point x,y approaches the maximum lethality distance, d_1, the chord length decreases to zero. In the calculation procedure described next, the minimum chord length is restricted to 5% of the lethality distance to produce a small but nonzero contribution as x,y approaches d_1.

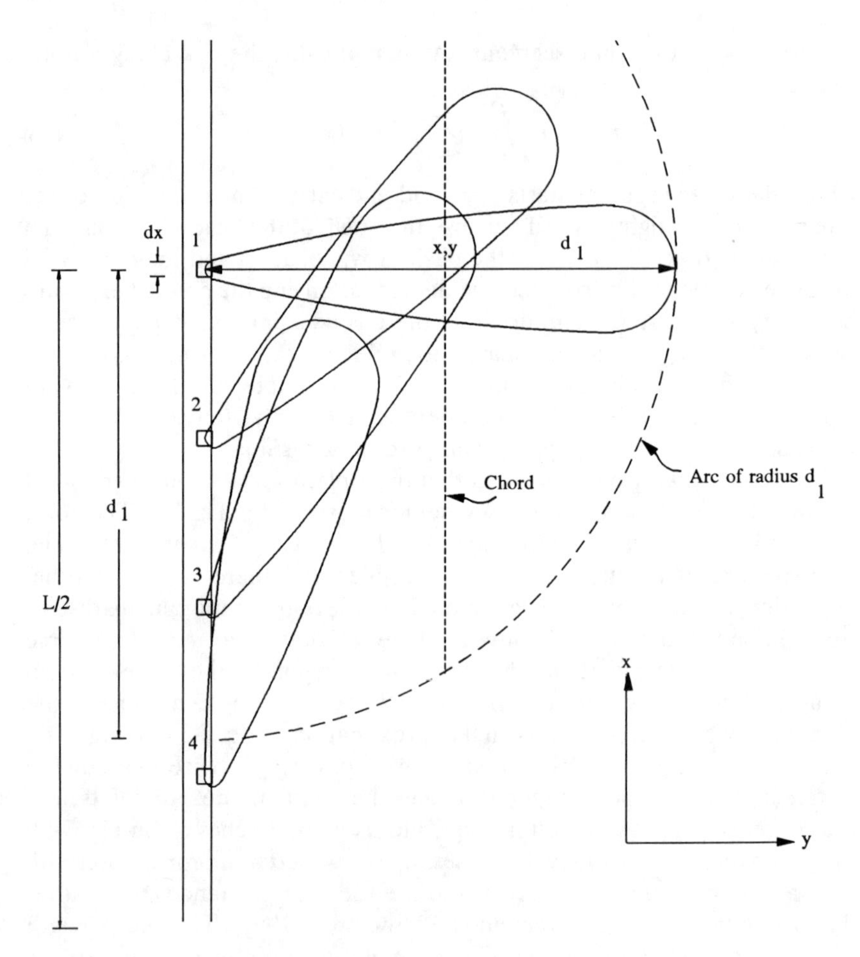

FIGURE 7-5. Contribution to the individual risk at *x,y* of one scenario occurring at several locations.

An assumption will be made to simplify the presentation of the procedure for calculating individual risk: a single isopleth for fatality will be used rather than the more complex procedure illustrated in Fig. 6-6. This commonly used assumption means that the probability of a fatality is 1 inside the isopleth and zero outside. The first step is to pick an arc width representing a wind direction. As a minimum, the consequence usually depends on which side of the road or the track the individual is located on; that is, the wind direction is subdivided into at least two 180-degree arcs.

A worksheet like the one in Table 7-1 should be used for the calculation. The scenarios are listed in the first column, and the values from Eq. 7-1 are inserted in the next five columns. An exception is the probability of a

TABLE 7-1. Individual risk calculation worksheet no. 1

Case Identification: Truck Shipment of Chlorine
Route Segment Identification: Residential Segment
Comments: LD_{50} distance is calculated; exposure inside contour is considered lethal; exposure outside is not lethal.

Scenario identification	Release frequency, yr^{-1}	Outcome probability	Evasive action probability	Weather probability	Health effect probability	Arc probability	Total frequency, yr^{-1}	Lethal distance, m
Impact, catastrophic failure, shelter, Cat. D, 2 m/sec (I1-1)	5.0×10^{-8}	0.6	0.95	0.5	1.0	0.5	7.1×10^{-9}	500
Impact, 6-in. hole, shelter, Cat. D, 2 m/sec (I1-2)	5.0×10^{-8}	0.4	0.95	0.5	1.0	0.5	4.8×10^{-9}	350
Impact, catastrophic failure, no shelter, Cat. D, 2 m/sec (I1-3)	5.0×10^{-8}	0.6	0.05	0.5	1.0	0.5	3.8×10^{-10}	1000
Impact, 6-in. hole, no shelter, Cat. D, 2 m/sec (I1-4)	5.0×10^{-8}	0.4	0.05	0.5	1.0	0.5	2.5×10^{-10}	800

TABLE 7-2. Individual risk calculation worksheet no. 2

Case Identification: Truck Shipment of Chlorine
Route Segment Identification: Residential Segment
Comments:

		Lethal Distance, m							
		350		500		800		1000	
Scenario identification	Scenario frequency, yr^{-1}	Chord correction	Modified frequency	Chord correction	Modified frequency	Chord correction	Modified frequency	Chord correction	Modified frequency
I1-3	3.8×10^{-10}	5.8×10^{-2}	2.2×10^{-11}	5.4×10^{-2}	2.0×10^{-11}	3.7×10^{-2}	1.4×10^{-11}	1.6×10^{-3}	6.1×10^{-13}
I1-4	2.5×10^{-10}	4.5×10^{-2}	1.1×10^{-11}	3.9×10^{-2}	9.8×10^{-12}	1.2×10^{-3}	3.0×10^{-13}		
I1-1	7.1×10^{-9}	2.2×10^{-2}	1.6×10^{-10}	7.8×10^{-4}	5.5×10^{-12}				
I1-2	4.8×10^{-9}	5.4×10^{-4}	2.6×10^{-12}						
TOTAL			2.0×10^{-10}		3.5×10^{-11}		1.4×10^{-11}		6.1×10^{-13}

population being present, $P_{i,p}$, as the individual risk calculation is based on the assumption that people are present. The release frequency is computed on the basis of the total segment length with the chord correction factor applied in Table 7-2. The sixth column is the probability that the wind is blowing into the arc chosen; in this case the probability is 0.5 for a 180-degree arc if the wind distribution is uniform. The formulas from Chapter 6 are used to determine the maximum lethal distance for each of the scenarios. (The values in Table 7-1 are for illustrative purposes only.)

The second step in the calculational procedure is shown in Table 7-2. The scenarios are listed in the first column, the scenario producing the largest lethal distance first, the scenario producing the second largest lethal distance second, and so forth. In the second column the total frequency from Table 7-1 is listed. Across the top of the next four columns the lethal distances from Table 7-1 are entered. The segment length in this example is 32,180 m (20 miles). For the I1-3 scenario, the correction factor at the maximum distance (1000 m) is 5% of 1000 divided by the segment length, or $50\ m/3.2 \times 10^4\ m = 1.6 \times 10^{-3}$. At 800 m, the chord for a circle of radius 1000 m is $(2)(1000)(1 - .8^2)^{1/2} = 1200$ m, and the chord correction factor is $1200\ m/3.2 \times 10^4\ m = 3.7 \times 10^{-2}$. Because the frequency in column two applies up to the maximum lethal distance, the rows for the modified frequency at each distance can be filled in after the chord correction factor at each distance is computed. The sum of the columns gives the frequency of a fatality at the distances indicated. Individual risk contours or a risk profile can be constructed using the values in Table 7-2. The plot of the risk contours from Table 7-2 will look like a series of lines parallel to the transport route. (See Fig. 7-2.) The risk profile from values in Table 7-2 is shown in Fig. 7-3. Additional distance points can be chosen to smooth or extend the curve in Fig. 7-3.

7.2.2 Societal Risk Calculation

The societal risk calculation is the same as the individual risk calculation up to and including determination of the concentration isopleths. An additional step is to determine the number of persons affected. The risk equation is:

$$R_s = \sum_{i=1}^{i_{max}} F_i, N_i \tag{7-5}$$

$$N_i = P_i P_{i,h} \tag{7-6}$$

where:

N_i = the number of people affected by scenario i
P_i = the number of people within the concentration isopleth
$F_i, P_{i,h}$ = parameters from Eq. 7-1

The first step in the calculation is to choose the number of wind directions to be used if the consequences depend on wind direction. Eight wind directions are illustrated in Fig. 7-6 when symmetry is considered. The choice of fewer or more wind directions depends on whether additional accuracy is worth the additional cost. For the case shown in Fig. 7-6, the plume is long enough to affect multiple population zones; therefore, eight wind directions are used to improve the accuracy. Note that the top two

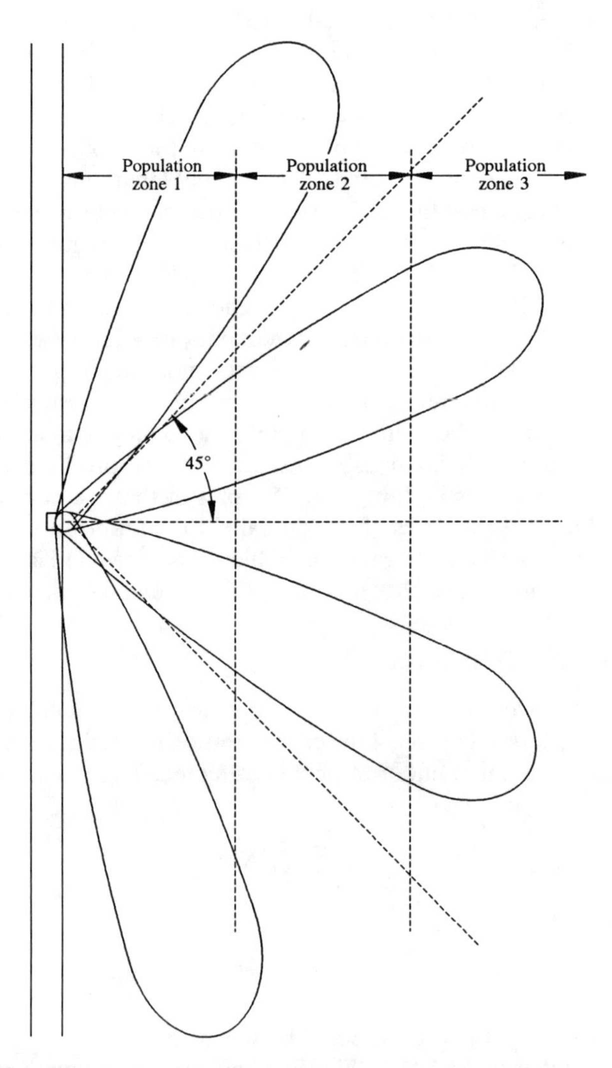

FIGURE 7-6. Contribution to societal risk with multiple wind directions and multiple population boundaries.

TABLE 7-3. Cumulative frequency worksheet

Bin limit	Frequency	Cumulative frequency
1-2	1.2×10^{-5}	3.1×10^{-5}
2-5	1.2×10^{-5}	1.9×10^{-5}
5-10	8.9×10^{-7}	7.0×10^{-6}
10-20	8.0×10^{-7}	6.0×10^{-6}
20-50	3.8×10^{-6}	5.3×10^{-6}
50-100	2.9×10^{-7}	1.5×10^{-6}
100-200	2.2×10^{-7}	1.2×10^{-6}
200-500	9.1×10^{-7}	9.8×10^{-7}
500-1000	7.2×10^{-8}	7.2×10^{-8}

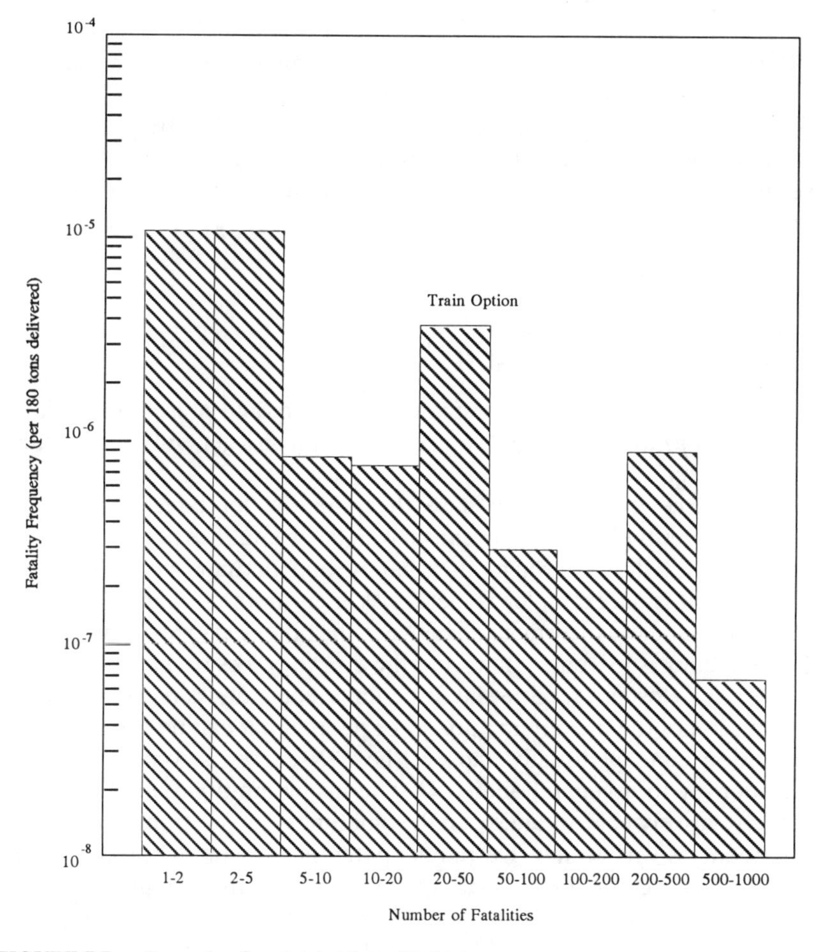

FIGURE 7-7. Example of societal risk profile histogram.

plumes are mirror images of the bottom two, and calculations based on Fig. 7-6 can be limited to two plumes and multiplied by two. Symmetry on the other side of the transportation route also may be involved. Because the population zones are bands of constant population density along the entire route segment, the plumes from an accident at another location along the route segment would produce the same result.

The number of scenarios that result from the analysis can be large, and it is unlikely that the number of people affected by two different scenarios will be the same. That is, if there are 175 scenarios, there are probably 175 unique values of people affected. To sum these values (Eq. 7-5), the results are grouped into ranges of values as shown in Table 7-3. This process also is called binning, as the frequency values for a specified range of consequences are accumulated in a bin.

Figure 7-7 presents the results of Table 7-3 in a histogram format. The results also could be presented in graphical form, but the large variation from bin to bin would appear to detract from a graphical presentation.

Societal risks usually are presented as an *F-N* curve. The frequency, *F*, is the frequency of an event with consequences affecting *N* or more persons. An example *F-N* curve is shown in Fig. 7-4. Table 7-3 can be used to compute the values for the curve in the figure.

7.3 SELECTION OF RISK PRESENTATION APPROACH

Experts agree that there is no best way to measure and present the risk of transporting hazardous materials by truck or train, on a general level or often even for a specific situation.

People perceive events that have the potential for large numbers of fatalities differently from those events in which few fatalities occur. Disasters cause grief and shock throughout a nation and perhaps throughout the world; Bhopal is an example. Societal risk measurement responds to this tendency by a society to distinguish between the potential for a disaster (even if the frequency is extremely low) and the everyday loss of many lives with only one or two lost in a single event. Inherent in societal risk measurement is an estimation of the maximum number of people that can be harmed. The presentation concept usually chosen, the *F-N* curve, is sometimes difficult to convey to a general audience. The histogram approach (Fig. 7-7), which is relatively easy to understand, sometimes "looks strange" if bin-to-bin frequencies vary widely. Despite its potential conceptual problem, the *F-N* curve has become the norm for quantitative risk analyses and the comparison of risk alternatives for both stationary plants and transportation routes.

Both societal risk and individual risk can be computed and presented for each route segment to enhance understanding of risk variation along the route.

Because of its geographic nature, individual risk is probably of more interest to persons who work or live along a particular route segment than is societal risk; however, individual risk profiles or contours generally reflect numerical values that are so small that they may not be meaningful to the public. Although individual risk values may be small and thus acceptable, the societal risk values may be much more significant and potentially unacceptable. Individual risk may not reflect the true risk to an individual, owing to vulnerability variations in the populations (Section 6.4).

Individual risk is not proportional to the route length, but societal risk is. Some (Anthony and Peirson 1991) argue that individual risk provides a measure of equitable distribution of risk, whereas societal risk is more comprehensive in the scope of the factors included. The inclusion of more factors probably makes societal risk better than individual risk for comparison of different routing options.

Another risk presentation approach is the risk index. For example, risk as the product of frequency and consequence is an index; and there are other indices. However, indices are not generally useful because they suffer from both ambiguity and problems of balance. An index is ambiguous because it is not easy to understand why the index value is high or low. Is the frequency high, or is the consequence high? A high consequence can be balanced by a low frequency; therefore, if an index is neither high or low, what can one conclude?

It is recommended that transportation quantitative risk analyses show results in the *F-N* curve format, except in unusual situations, because of the universality of their use. Individual risk curves may be useful in some cases. The question of the acceptability of the calculated risk, that is, risk assessment, is a sociopolitical issue outside the scope of this book.

The presentation of the risk triplets should accompany the *F-N* curve format. These triplets show the dominant contributors to risk, and thus help to answer the question of where to apply resources to mitigate the risk.

References

Anthony, D. K., and J. Peirson, Jr. 1991. Managing risks of hazardous materials transportation in Santa Barbara County. In *State and Local Issues in Transportation of Hazardous Waste Materials: Towards a National Strategy,* ed. M. D. Abkowitz and K. G. Zografos,. pp. 120-133. New York: American Society of Civil Engineers.

Saccomanno, F. F., and J. H. Shortreed. 1991. Societal-individual risks for hazmat transportation. In *State and Local Issues in Transportation of Hazardous Waste Materials: Towards a National Strategy,* ed. M. D. Abkowitz and K. G. Zografos, pp. 220-245. New York: American Society of Civil Engineers.

8

Example: Bulk Transport of Chlorine by Truck and Train

This chapter presents a detailed calculation of the risk of bulk transport of chlorine by truck and by train. The purpose of the example is to demonstrate the features unique to transportation QRA: development of transportation accident scenarios, use of transportation databases, and segmentation of the route(s) producing multiple accident locations. As with any analysis, various facets could be considered in more detail. For example, only one route segment is used to characterize each of the urban, suburban, and rural portions of the postulated routes. The addition of more segments to the examples would increase the mathematical complexity without adding to the reader's understanding of transportation QRA. This same philosophy has been used throughout this illustrative example: completeness but not necessarily complexity.

This book and this example cover accidents that occur during the transport phase, including short-term storage in parking lots or railyards. Loading, unloading, and long-term storage are not addressed. Loading and unloading risks may be much more significant than those from the actual transport phase (Kirchner and Rhyne 1992). Loading and unloading risk is not covered in this book because the analysis involves different QRA techniques and databases from those presented here.

A major objective of this chapter is to provide the details behind the presentation of risk results. The societal risk concept frequently is difficult to understand; the individual risk concept also is sometimes confusing. The most technical information is communicated by a triplet: accident scenario, frequency, and consequence. All three presentation techniques will be developed in detail.

Disclaimer: This example is for illustrative purposes only. The data do not represent any real situation. The conclusions about the risk of highway transport, the risk of rail transport, and the relative risk between highway and rail are appropriate only for this example and cannot be generalized to other situations with chlorine or another hazardous material.

8.1 THE DECISION

The ACME Processing Company has built a new plant near the city of Green Valley, and, as part of its commitment to the city council of Green Valley, ACME has promised to perform a quantitative risk analysis of transporting hazardous materials to and from the facility. The ACME process uses large amounts of chlorine, and the nearby Chlorine Company is the likely source. All parties have agreed to let the results of the risk analysis determine whether the transportation mode used will be truck or train because the net effect of all other factors influencing the choice result in not a strong recommendation for either mode. Three combinations of route and mode choices appear feasible, as shown in Fig. 8-1:

1. A relatively short highway route on US 40 through the center of Green Valley.

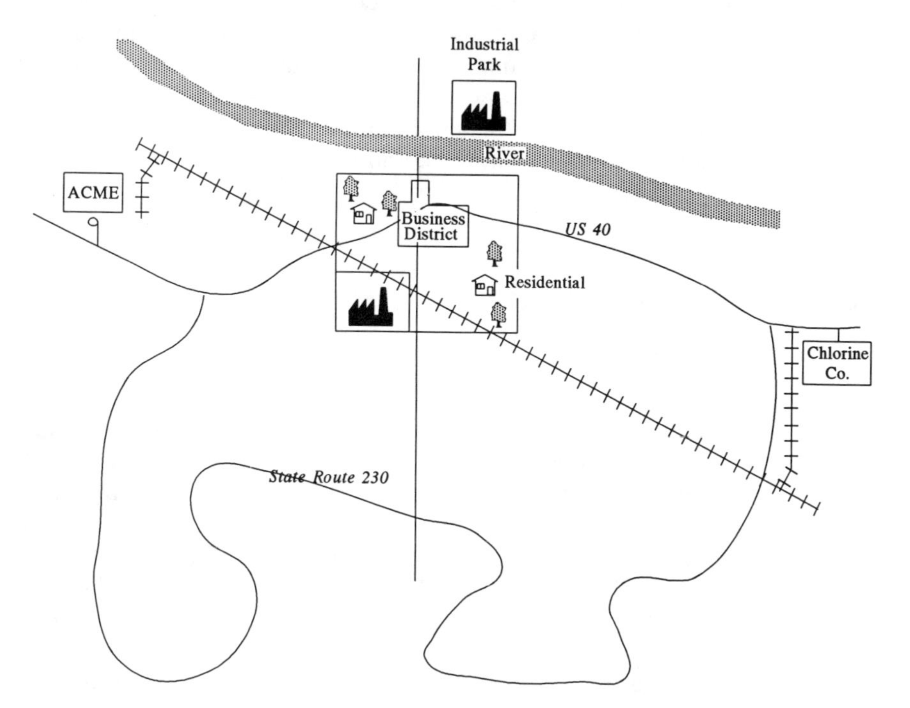

FIGURE 8-1. Route and mode alternatives.

TABLE 8-1. Data for alternative mode and route analysis

Parameter	Rail route	Highway US 40	Highway SR 230
Route characteristics	Class 3 track	2-lane	2-lane
Route length (miles)			
Rural	100	75	200
Residential (suburban)	20	20	—
Industrial/business (urban)	5	5	—
Total	125	100	200
Amount of chlorine per shipment (tons)	90	20	20

2. A longer highway route on State Route (SR) 230 through a low-population area, bypassing the city entirely.
3. A direct-rail route through a residential area and an industrial park within the city limits of Green Valley.

The basic data for these three choices are presented in Table 8-1.

The primary decision is to determine which route/mode choice presents the least risk to the citizens of Green Valley. Often secondary decisions arise once the primary decision is defined, as described in the following section. The reader will understand that any route that bypasses the town will result in essentially no risk to the town. A real situation will be more complex than this; each choice will have its own cost and will affect a unique group of people.

8.2 PRELIMINARY HAZARDS ANALYSIS

8.2.1 Objectives and Scope of the Analysis

The original objective for the analysis was to satisfy the officials of Green Valley, who specifically stated that the citizens were concerned about "what could happen," and the fire and police chiefs were concerned that "they might not be prepared for an emergency." These comments imply that the analysis should focus on the more severe, or bounding, events. ACME management wants the analysis to be responsive to the Green Valley officials, but does not want an elaborate analysis. As long as they are sponsoring an analysis, however, they are interested in identifying options that will reduce corporate financial risk from the transport of chlorine. (Risk reduction, clearly a primary reason for performing a risk analysis, frequently starts as a secondary objective or decision in real situations.)

The analysis objectives can be met with an analysis having the following scope:

1. Loading and unloading accidents are not to be addressed because both the Chlorine Company and ACME are too far from Green Valley for at-plant accidents to have an effect.
2. The potential for injuries is important, but an analysis based on fatalities is considered sufficient for management to choose among the mode and route alternatives.
3. Presentation of results will include both societal and individual risk. Societal risk will show emergency response needs and be sufficient for management to choose among the mode and route alternatives. Individual risk results will address some risk acceptability issues.

8.2.2 Hazards Identification

The hazard of interest is the chlorine being transported. The potential for toxic exposure is the concern. The potential for injuries or fatalities from a heavy truck accident is not, of itself, of interest for the specified decision.

8.2.3 Consequences of Interest

Although injuries from exposure to chlorine are of concern to ACME, the Chlorine Company, and the officials of Green Valley, the consequence that is expected to determine the mode/route choice is the potential for fatalities from exposure to chlorine.

8.2.4 Initiating Events

Chlorine exposure that may produce a fatality from either an accident or a nonaccidental release are of interest for the specified decision.

8.3 ACCIDENT SCENARIO DEVELOPMENT

The fault tree developed in Chapter 4 is based on a chlorine railway tank car. The list of release scenarios that can be identified from the fault tree for accidents and nonaccidents is presented in Fig. 8-2. The event tree analysis indicates that fires can follow releases from mechanical failure. These fires can affect release and dispersion characteristics; these separate scenarios are indicated in parentheses in Fig. 8-2. A total of 40 release scenarios arises for this analysis of the chlorine railcar: one incident (nonaccident), 13

Scenario Identifier	Scenario Description
RI	Nonaccident release, e.g., unintentional relief valve operation
ILV (ILVF)[1]	Impact fails liquid valves and excess flow valve does not close
IGV (IGVF)	Impact fails gas valves
ISD (ISDF)	Impact fails defective tank shell
IS (ISF)	Impact fails tank shell
IHD (IHDF)	Impact fails defective tank head
IH (IHF)	Impact fails tank head
IMD (IMDF)	Impact fails defective manway
IM (IMF)	Impact fails manway
PH (PHF)	Puncture probe fails tank head
PS (PSF)	Puncture probe fails tank shell
PV (PVF)	Puncture probe fails valve dome
CD (CDF)	Crush fails defective tank
C (CF)	Crush fails tank
FRV	Fire raises pressure to relief valve setpoint; tank is upright and less than 50% released; vapor release occurs
FRL	Fire raises pressure to relief valve setpoint; tank is overturned but fire duration does not overpressurize; liquid release occurs
FFF	Fire raises pressure above relief valve set point; relief valve fails to open; tank fails
FFL	Fire raises pressure above relief valve setpoint; tank is overturned and pressurization rate exceeds relief valve capability
FFI	Fire produces failure of upright tank when greater than 50% inventory loss occurs
FD	Fire produces failure of defective tank when relief pressure reached
FW	Fire produces failure of weakened tank when relief pressure reached
FWI	Fire produces failure of weakened tank with insulation damage when relief pressure reached
FRVI	Fire failure scenario FRV preceded by insulation damage
FRLI	Fire failure scenario FRL preceded by insulation damage
FFFI	Fire failure scenario FFF preceded by insulation damage
FFLI	Fire failure scenario FFL preceded by insulation damage
FFII	Fire failure scenario FFI preceded by insulation damage

[1] Parenthesis indicates a separate scenario in which fire occurs after mechanical forces cause container failure.

FIGURE 8-2. Release scenarios.

mechanical failures, 13 mechanical failures followed by fire, five fire failures of normal tanks, five fire failures of normal tanks with insulation damage, and three failures of defective or weakened tanks at or before the relief pressure is reached. Tank car orientation is considered explicitly in several scenarios and will affect others at the consequence stage (Section 8.5). Accidents producing mechanical failure can result in a hole in the vapor space or in the liquid space. A small hole in the vapor space will result in vapor release, but a large hole will result in aerosol entrainment in the vapor. The result of these variations is that the number of scenarios increases to more than 50. The chlorine tank truck has an analogous list of accident scenarios; thus, the total number of scenarios to be evaluated for frequency and consequence is about 100. The analyst will want to look continually for ways to group accident scenarios together because there are several factors yet to be discussed (e.g., route segments and meteorology) that will increase the number of calculations even more.

8.4 FREQUENCY ANALYSIS

The initiating event for all but the nonaccident scenario is a truck or train accident that usually is expressed as a frequency, that is, number of accidents per mile, per year, or per quantity transported. Other frequency-related factors evaluated in this section are conditional probabilities for accident force type, force magnitude, and release modes. Probabilities arising from the consequence calculations will be considered when those calculations are made.

The first step is to divide each route into segments along which all characteristics needed to calculate both the frequency and the consequence can be averaged without introducing large errors. The major considerations are (from Fig. 2-3) as follows: accident frequency, which depends on road or track characteristics; dispersion, which depends on meteorology and terrain; and population density. For State Route 230, the entire route is a two-lane road through a rural area; thus, one route segment (i.e., the entire route) is appropriate. US Route 40 is also entirely a two-lane road, but three distinct population groups are encountered: rural, suburban, and urban. Three route segments are needed for population calculations and perhaps for characterization of the highway accident rate. The railway track class is uniform; therefore, only one route segment is needed from an accident rate viewpoint, but three distinct population groups are needed: rural, suburban, and urban.

8.4.1 Accident Initiator Frequency

Of the three options described in Section 3.1.1 to generate a highway accident rate for trucks, the most expedient one is to use a value from Table 3-4 or 3-5. The descriptors in Table 3-4 are the more applicable, so values from it will be used here.

State Route 230

$$\left(200\ \frac{\text{rural miles}}{\text{shipment}}\right)\left(\frac{2.19\times10^{-6}\ \text{accidents}}{\text{rural mile}}\right) = 4.38\times10^{-4}\ \text{accidents/shipment}$$

$$\left(\frac{9\ \text{shipments}}{180\ \text{tons}}\right)\left(\frac{4.38\times10^{-4}\ \text{accidents}}{\text{shipment}}\right) = 3.94\times10^{-3}\ \text{accidents/180 tons delivered}$$

US 40

$$\left(\frac{5\ \text{urban miles}}{\text{shipment}}\right)\left(\frac{8.66\times10^{-6}\ \text{accidents}}{\text{suburban mile}}\right) = 4.33\times10^{-5}\ \text{accidents/shipment}$$

$$\left(\frac{20\ \text{suburban miles}}{\text{shipment}}\right)\left(\frac{8.66\times10^{-6}\ \text{accidents}}{\text{urban mile}}\right) = 1.73\times10^{-4}\ \text{accidents/shipment}$$

$$\left(\frac{75\ \text{rural miles}}{\text{shipment}}\right)\left(\frac{2.19\times10^{-6}\ \text{accidents}}{\text{rural mile}}\right) = 1.64\times10^{-4}\ \text{accidents/shipment}$$

$$3.80\times10^{-4}\ \text{accidents/shipment}$$

$$\left(\frac{9\ \text{shipments}}{180\ \text{tons}}\right)\left(\frac{3.80\times10^{-4}\ \text{accidents}}{\text{shipment}}\right) = 3.42\times10^{-3}\ \text{accidents/180 tons delivered}$$

Rail

$$\left(\frac{125\ \text{miles}}{\text{shipment}}\right)\left(\frac{4.7\times10^{-6}\ \text{accidents}}{\text{train mile}}\right)(2.2\ \text{class 3 factor}) = 1.29\times10^{-3}\ \text{train accidents/shipment}$$

$$\left(\frac{2\ \text{shipments}}{180\ \text{tons}}\right)\left(1.29\times10^{-3}\ \frac{\text{accidents}}{\text{shipment}}\right) = 2.58\times10^{-3}\ \frac{\text{train accidents}}{\text{180 tons delivered}}$$

FIGURE 8-3. Accident initiator frequency calculations.

8.4.1.1 State Route 230

The accident rate from Table 3-4 for a rural, two-lane road is 2.19×10^{-6} accidents per mile, and the accident frequency per shipment for the example is given in Fig. 8-3. It is useful to compute an accident rate for an equal tonnage by truck and train, which would be nine trips by truck of 20 tons each and two trips by train of 90 tons each. For State Route 230 the accident frequency is 3.94×10^{-3} accidents/180 tons delivered.

8.4.1.2 US Route 40

Table 3-4 does not distinguish between urban and suburban two-lane roads, and we will use the same accident rate for both, 8.66×10^{-6} accidents/mile. Accident frequencies per shipment for each segment and per 180 tons delivered for the entire route are given in Fig. 8-3.

8.4.1.3 Railway

The train accident rate for the last six years shown in Fig. 3-1 was about 4.7×10^{-6} accidents/train mile. A class 3 track accident adjustment factor of 2.2 is obtained from Table 3-9. Accident frequencies per train shipment and per 180 tons delivered are given in Fig. 8-3.

8.4.2 Force Type, Force Magnitude, and Container Failure

The force magnitude probability distributions and the container failure threshold are evaluated together to produce a conditional probability of failure, given that the force type occurs. The datum for force type probability often is intertwined with the force magnitude probability, and it usually is convenient to evaluate these components together. The numerical evaluation of these components is presented in detail in Appendix A. The results of combining the accident initiator frequencies with the conditional probabilities for force type, force magnitude, and container failure are given in Table 8-2.

8.4.2.1 Train Failure Type Distribution Comparison with Data

Accident data over the time span 1965 through 1986 are given in Fig. 3-18 for several types of tank cars. The number of losses for a specific type of tank car divided by the total number of tank car losses results in a distribution of the type of loss. The first three columns of Table 8-3 show the distribution of failure types for: all pressure cars; for 105A and 120A cars; and for 112 (S, J, T), 114 (S, J, T), and 105 (S, J) cars, respectively. The fourth column shows the results of the preceding frequency analysis divided by the total accident

TABLE 8-2. Frequency analysis summary

	Frequency[a]	
Scenario identifier	Train	Truck (SR 230)
RI	2.0×10^{-4}[b]	6.4×10^{-5}
ILV (ILVF)	2.4×10^{-8} (2.4×10^{-10})	2.5×10^{-8} (6.0×10^{-10})
IGV (IGVF)	1.1×10^{-5} (1.1×10^{-7})	1.1×10^{-5} (2.6×10^{-7})
ISD (ISDF)	4.7×10^{-8} (4.7×10^{-10})	4.4×10^{-8} (1.0×10^{-9})
IS (ISF)	6.7×10^{-6} (6.7×10^{-8})	3.7×10^{-6} (8.9×10^{-8})
IHD (IHDF)	9.0×10^{-8} (9.0×10^{-10})	3.1×10^{-7} (7.4×10^{-9})
IH (IHF)	3.0×10^{-6} (3.0×10^{-8})	2.4×10^{-6} (5.8×10^{-8})
IMD (IMDF)	6.3×10^{-9} (6.3×10^{-11})	5.9×10^{-9} (1.4×10^{-10})
IM (IMF)	9.0×10^{-7} (9.0×10^{-9})	4.9×10^{-7} (1.2×10^{-8})
PH (PHF)	7.2×10^{-7} (7.2×10^{-9})	5.7×10^{-6} (1.4×10^{-7})
PS (PSF)	5.4×10^{-7} (5.4×10^{-9})	9.2×10^{-7} (2.2×10^{-8})
PV (PVF)	2.1×10^{-9} (2.1×10^{-11})	7.0×10^{-9} (1.7×10^{-10})
CD (CDF)	4.4×10^{-9} (4.4×10^{-11})	0.0 (0.0)
C (CF)	4.1×10^{-7} (4.1×10^{-9})	0.0 (0.0)
FRV	4.0×10^{-6}	1.1×10^{-6}
FRL	9.7×10^{-7}	1.1×10^{-7}
FFF	1.3×10^{-9}	2.6×10^{-10}
FFL	9.7×10^{-7}	2.1×10^{-8}
FFI	3.6×10^{-7}	6.2×10^{-8}
FD	1.6×10^{-8}	3.3×10^{-9}
FW	0.0	0.0
FWI	1.4×10^{-7}	4.4×10^{-7}
FRVI	4.3×10^{-7}	1.7×10^{-7}
FRLI	1.5×10^{-7}	2.0×10^{-7}
FFFI	1.8×10^{-10}	4.7×10^{-10}
FFLI	1.2×10^{-7}	1.1×10^{-7}
FFII	1.8×10^{-7}	2.3×10^{-7}

[a]Frequency per 180 tons delivered.
[b]Value reduced to 2×10^{-5} in Appendix B.

frequency to obtain a conditional failure probability. The analysis for the 105A500W chlorine tank car in the example is least comparable to the first column.

The most pronounced difference between the columns is the calculated impact rupture, which is an order of magnitude higher than the corresponding value for any of the previous three columns. This indicates that the calculated impact failure threshold is too conservative (too low), the impact force magnitude model is too conservative (too high), or both. In this analysis, both are probably overly conservative. All other analytical values are within an order of magnitude of the accident data, and generally within a

TABLE 8-3. Comparison of conditional failure probability data and analysis

Cause of Loss	All pressure cars	105A 120A cars	105S,J 112 (114)S,J,T cars	Analysis results for 105A500W chlorine car
Head puncture	0.20	0.11	0.05	0.02
Shell puncture	0.13	0.16	0.16	0.02
Attachment damage	0.007	0.02	0	[b]
Bottom ftg's damage	0.02	0.02	0.01	[b]
Top ftg—mech damage	0.05	0.06	0.11	0.35
Top ftg—fire damage	0.02	0.02	0.08	[b]
Top ftg—no damage (loose)	0.16	0.22	0.21	[b]
Top ftg—unknown	0.07	0.07	0.06	[b]
Impact rupture	0.02	0.02	0.03	0.35
Fire rupture	0.15	0.08	0.10	0.06
Safety valve discharge[a]	0.08	0.14	0.12	0.18
More than one cause	0.06	0.04	0.06	[b]
Unknown cause	0.04	0.04	0	[b]
Crush rupture	[b]	[b]	[b]	0.01
	1.00	1.00	0.99	0.99
Total Failures	731	202	89	N/A

[a]Valve not damaged, opened owing to fire or mechanical damage elsewhere.
[b]Not evaluated explicitly, in some cases because not applicable to 105A500W car.

much smaller range. The mechanical damage to valves loss category, listed as "Top ftg—mech damage," consists essentially 100% of the IGV scenario, and the relatively large value for the example analysis indicates that the fraction of impacts that affect the valve dome and/or the impact failure threshold used in the analysis of any nonzero value also is too conservative. The low puncture values are not unexpected because the 105A500W car is relatively more puncture-resistant than the average car in any of the first three columns. The 105A500W car also is more thermally resistant than the average railcar in any of the first three columns. Therefore, the high "fire rupture" and "safety valve discharge" values indicate that the calculated fire thresholds and/or the thermal force magnitude model is probably conservative. In general, the comparison of the analytical results with the data is remarkably good. (The loose top fitting release scenario is not in the fault tree. The release mechanism is significant from a frequency standpoint but not from a release standpoint, as generally only a few gallons are released in these events.)

The contribution of the individual scenario results within the broad damage categories is shown in Table 8-4. A few broad generalizations can be

TABLE 8-4. Grouping of train accident scenarios into failure categories

Failure category	Scenario identifier	Accident frequency per 180 tons delivered	Percentage contribution to total
Impact rupture of tank	ISD	4.7×10^{-8}	
	ISDF	4.7×10^{-10}	
	IS	6.7×10^{-6}	62
	ISF	6.7×10^{-8}	
	IHD	9.0×10^{-8}	
	IHDF	9.0×10^{-10}	
	IH	3.0×10^{-6}	28
	IHF	3.0×10^{-8}	
	IMD	6.3×10^{-9}	
	IMDF	6.3×10^{-11}	
	IM	9.0×10^{-7}	8
	IMF	9.0×10^{-9}	
		1.1×10^{-5}	98
Mechanical damage to valves	ILV	2.4×10^{-8}	
	ILVF	2.4×10^{-10}	
	IGV	1.1×10^{-5}	100
	IGVF	1.1×10^{-7}	
	PV	2.1×10^{9}	
	PVF	2.1×10^{-11}	
		1.1×10^{-5}	100
Crush rupture of tank	CD	4.4×10^{-9}	
	CDF	4.4×10^{-11}	
	C	4.1×10^{-7}	99
	CF	4.1×10^{-9}	
		4.2×10^{-7}	99
Head puncture	PH	7.2×10^{-7}	99
	PHF	7.2×10^{-9}	1
		7.3×10^{-7}	100
Shell puncture	PS	5.4×10^{-7}	99
	PSF	5.4×10^{-9}	1
		5.5×10^{-7}	100
Fire lifts valve	FRV	4.0×10^{-6}	72
	FRL	9.7×10^{-7}	17
	FRVI	4.3×10^{-7}	8
	FRLI	1.5×10^{-7}	3
		5.6×10^{-6}	100

Continued

TABLE 8-4. ***Continued***

Failure category	Scenario identifier	Accident frequency per 180 tons delivered	Percentage contribution to total
Fire rupture	FFF	1.3×10^{-9}	
	FFL	9.7×10^{-7}	54
	FFI	3.6×10^{-7}	20
	FD	1.6×10^{-8}	1
	FW	0.0	
	FWI	1.4×10^{-7}	9
	FFFI	1.8×10^{-10}	
	FFLI	1.2×10^{-7}	7
	FFII	1.8×10^{-7}	9
		1.8×10^{-6}	100

made from the table. Scenarios involving a defective component are negligible compared to the normal component because defective component probabilities are smaller by two orders of magnitude than the increase in failure probability owing to the lower failure threshold. The frequency contribution of accident scenarios involving fire following container failure from mechanical forces is negligible compared with the mechanical failure without fire scenario. In this case, however, the consequences differ and cannot be discarded from the risk analysis on the basis of frequency values alone. These general conclusions also apply to the truck frequency analysis.

8.4.3 Incident Initiator Frequency

Chlorine railcars have a leak rate of approximately 1/10,000 cars shipped (Section 3.3.3). For the two rail trips of interest, the leak frequency is 2.0×10^{-4} leaks/180 tons delivered. As shown in Table 3-17, the number of highway hazmat incidents involving valve or fitting failure (3,289) is 2.3 times the number of traffic accident releases (1,427). The sum of the truck accident releases in Table 8-2 is 2.76×10^{-5}/180 tons delivered for SR 230; thus, the tank truck leak frequency is 6.35×10^{-5} leaks/180 tons delivered for SR 230. These values are entered into Table 8-2 for the RI scenario.

8.5 CONSEQUENCE ANALYSIS

The frequencies of 40 release scenarios for train and a like number for truck were evaluated in Section 8.4. The first step in consequence analysis is to group the scenarios with similar source terms. A source term is a description of the amount of a hazardous release (or a release rate) together with any other information needed to characterize the release (e.g., release temperature). The second step in consequence analysis is to quantify the exposure of persons to the source term and the health effect of the exposure, and the third step is to estimate the number of persons exposed, including their potential escape and other mitigating actions.

8.5.1 Characterization of the Source Terms

Each of the 40 release scenarios results in a source term that probably is somewhat unique. In many cases, however, the data needed to quantify the differences between scenarios are lacking. For example, how does a shell impact failure differ in the size of the hole produced or its location from a shell puncture? A second factor to consider is orientation of the tank car with respect to the postulated hole in the tank car. If the hole is on the top and the tank car is upright (No. 2 in Fig. 6-2), then either a vapor release or a vapor plus aerosol release occurs; but if the hole is below the liquid level (No. 3 in Fig. 6-2), then a superheated liquid release occurs. The releases will vary both in the discharge rate and in the total amount of the discharge.

The release scenarios listed in Fig. 8-2 are examined in Appendix B to characterize the source term of each and to group scenarios with similar source terms. The initial characterization of the 40 train accident scenarios shows that when the potential hole location, hole size, and tank car orientation are considered, many of the scenarios have multiple release characteristics that must be evaluated as if they were multiple scenarios. At the same time, it is apparent that some scenarios can be grouped together because the consequences will be similar. The net result is that the 40 original train accident scenarios are reduced to 23. Each scenario applies to both truck and train; thus, 46 scenarios must be evaluated. This is too costly for practical decision making and too lengthy and complicated for this example. The 23 scenarios are reviewed and grouped into five scenarios as described in Appendix B. The resulting five source terms are presented in Table 8-5. The RI and IGV releases involve valves that are the same for both truck and train, as described in Appendix B. The IS, FRV, and FWI releases are different for the truck and the train modes. Thus, there are effectively eight different source terms to be evaluated.

8.5.2 Exposure and Response Quantification

After determining the characteristics of the source term, the analyst next estimates the extent to which people are exposed to the source term and the

TABLE 8-5. Release characteristics for release scenarios

Applicable scenario identifier	Release characteristics
	Initial Conditions
	Ambient temperature: 25°C (77°F)
	Chlorine temperature: 25°C (77°F)
	Chlorine pressure: 7.70×10^5 N/m^2 (97 psig)
	Liquid chlorine density: 1392 kg/m^3 (86.9 lb/ft^3)
	Valve Releases
IGV, RI	Relief valve flow area: 5.06×10^{-4} m^2 (5.45×10^{-3} ft^2)
IGV	Initial liquid flow rate at 25°C (Eq. 6-5): 13 kg/s (28.6 lb/s)
IGV, IS	Temperature after flashing: −34°C (−29°F)
RI	Initial vapor flow rate at 25°C (Eq. 6-2): 1.06 kg/s (2.34 lb/s)
FRV (train)	Initial vapor flow rate at 82°C (Eq. 6-2): 4.03 kg/s (8.89 lb/s)
FRV (truck)	Initial vapor flow rate at 60°C (Eq. 6-2): 2.60 kg/s (5.83 lb/s)
	Large Hole Release
IS	Flow area: 0.0324 m^2 (0.349 ft^2)
IS	Initial liquid flow rate at 25°C (Eq. 6-5): 830 kg/s (1830 lb/s)
IS (train)	Duration of release: 98.5 s
IS (truck)	Duration of release: 21.8 s
	Instantaneous Release at Elevated Temperature
FWI (train)	Complete inventory: 8.16×10^4 kg (90 short tons)
FWI (truck)	Complete inventory: 1.81×10^4 kg (20 short tons)
FWI (train)	Initial release temperature: 82°C (180°F)
FWI (train)	Effective release temperature: 127°C (261°F)
FWI (truck)	Initial release temperature: 60°C (140°F)
FWI (truck)	Effective release temperature: 105°C (221°F)

health effect of the exposure. For most toxic materials, exposure occurs during the downwind dispersion of a plume. Eight different source terms are identified in Table 8-5. The four with large release rates have release times varying from instantaneous (FWI truck and train) to 98.5 s (IS train). The distance downwind at which potentially harmful effects will occur will be shown to be on the order of a few thousand meters. At relatively large values of wind velocity, for example, 10 m/s (22.4 mph), the time to travel 2000 m is 200 s, which exceeds the release time; the travel time will be even longer for most wind speeds of interest, such as 1 to 5 m/s. The downwind plume will have the characteristics of a puff rather than a continuous release for these four releases. The remaining four releases are characterized by lower release rates and as a result have longer durations; these releases have characteristics of relatively continuous releases to people downwind. Modeling of the eight releases will require both puff and continuous plume calculations.

A major consideration is the selection of meteorological parameters for the calculation, that is, the atmospheric stability (Table 6-1) and the wind speed. The values of these two parameters will affect the calculated downwind concentration by orders of magnitude. The normal procedure is to use a minimum of two or three sets of meteorological parameters, each having an appropriate conditional probability, and perhaps many more sets. The dispersion from each release is evaluated by using each set of meteorological parameters. For this example, no useful purpose is served by making more than one set of dispersion calculations. To make the calculations as realistic

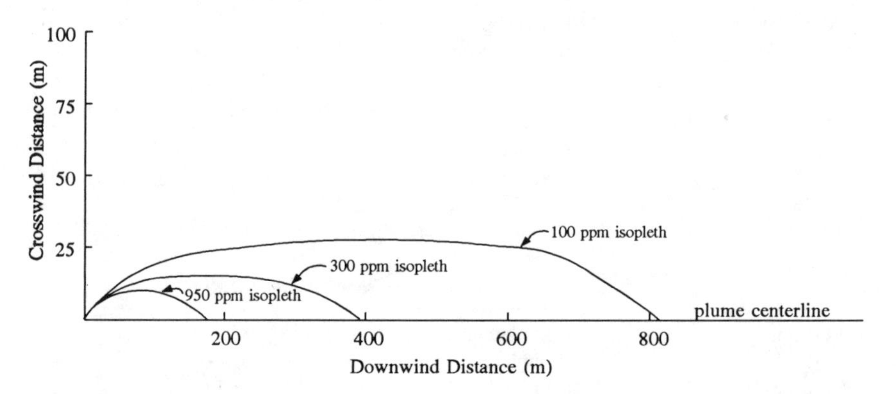

FIGURE 8-4. Typical continuous release calculation results showing three concentration isopleths.

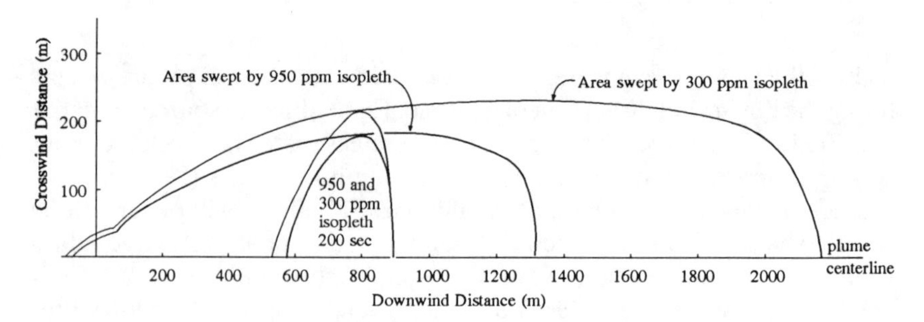

FIGURE 8-5. Typical puff release calculation results showing two concentration isopleths 200 s after release and the area swept by the two isopleths.

as possible, D stability and 5 m/s wind speed are chosen for the calculations. For many situations, this combination represents 50-50 meteorology; that is, the meteorology will be better 50% of the time and worse 50% of the time.

The downwind dispersion of the eight release cases is evaluated by using the DEGADIS heavy gas computer program (Trinity Consultants 1991). An example continuous release calculation is shown in Fig. 8-4, and an example puff release calculation is shown in Fig. 8-5. The parameters of interest are the areas within isopleths. Note that the crosswind scale is shown smaller than the downwind scale so that the width of the isopleth is exaggerated with respect to the length. The dimensions of the puff at 200 s after release are shown in Fig. 8-5 for two different concentrations. If a person were located at a point 800 m downwind, the puff would pass in about 80 s. Typically the puff passage time is less than 2 min for all cases at all points downwind. Note in Fig. 8-5 that the release spreads upwind a short distance. This occurs for both instantaneous release cases. The areas within the isopleths for the eight cases are given in Table 8-6. Dispersion calculations are not unique to transportation risk analysis; so the calculational procedure is not described in detail.

The truck and the train FWI release cases are representative of instantaneous releases. Owing to the effects of a large fire external to the tank car or tank truck, the temperature of the material released has been increased by 45°C to 50°C in Appendix B to conservatively account for heating of the material after the release and before dispersion away from the fire. The

TABLE 8-6. Exposure/response calculation results

Scenario ID	Area (km^2)	Isopleth concentration (ppm)
IGV	1.18×10^{-2}	950
	1.48×10^{-1}	100
RI	3.50×10^{-4}	950
	6.72×10^{-3}	100
FRV (train)	3.18×10^{-3}	950
	3.65×10^{-2}	100
FRV (truck)	1.32×10^{-3}	950
	1.85×10^{-2}	100
IS (train)	6.63×10^{-2}	16,000
	8.55×10^{-1}	950
IS (truck)	2.76×10^{-2}	16,000
	3.31×10^{-1}	950
FWI (train)	1.14×10^{-1}	16,000
	1.26×10^{0}	950
FWI (truck)	3.52×10^{-2}	16,000
	4.09×10^{-1}	950

version of DEGADIS used would not execute with the average temperature given in Table 8-5. Rather, the maximum temperature that could be used was 95°C. The effect of the high release temperature of 95°C was to give the release substantial buoyancy, but not enough to keep the release elevated for more than about 30 s. A large fire would create an additional updraft, which would contribute to the buoyancy of the release. More accurate modeling would have produced a plume elevated for a longer time above the heads of people potentially exposed. The effect would have been to increase dispersion and thereby lower the concentration before people were exposed. The elevation by a large fire of the 60 tons of chlorine released from the 1979 incident at Mississauga, Canada, is the probable reason why no one was killed by the large release (Marshall 1987). Thus, the FWI results given in Table 8-6 probably are very conservative.

The characteristics of the exposure determine the toxicity criteria. Each of the four large releases produces a puff of chlorine that moves downwind, and each puff will pass a given point in less than 2 min. The LC_{50} for a 2-min exposure is 950 ppm (Fig. 6-4). In this example, all persons inside the LC_{50} isopleth will be counted as fatalities, and all those outside will not (Section 6.5.1). If a person is indoors or rapidly goes indoors, Eqs. 6-19 and 6-20 are used to evaluate the indoor concentration. An outdoor concentration of 16,000 ppm will produce a peak indoor concentration of 446 ppm in an average home as a puff of 2-min duration passes. After another 9 min, the indoor concentration will decrease only to 393 ppm. At that time people are assumed to realize that the outdoor air is safe and either go outside or open windows to "air out" the house. The LC_{50} for an 11-min exposure is 400 ppm. Therefore, the 16,000 ppm outdoor concentration isopleth is used as the LC_{50} for people indoors.

The four small, long releases last from 0.5 hr to several hours. The plume that forms is hundreds of meters long but is only 10 to 40 m wide from the centerline outward at the widest point of the plume (see Fig. 8-4). At a normal walking speed of 1.8 m/s (4 mph), a person in the center of the plume could walk out of it in less than 6 to 22 s. Therefore, the 950 ppm value also is used for the LC_{50} for persons outdoors and exposed to a small, long plume. Persons indoors in an average house will experience 58% of the indoor concentration after 1 hr and 82% after 2 hr. If the persons indoors are healthy and realize that they can escape by moving crosswind for several seconds, the probability of escape before serious health effects occur is excellent. Another possibility is evacuation by emergency response personnel after an hour or two. To keep this analysis simple, an outdoor concentration of 100 ppm is used as the LC_{50} isopleth for persons indoors, and no escape is assumed during the first 2 hr. In urban and suburban settings, this assumption is expected to be very conservative.

8.5.3 Population Exposed

The population is characterized for this sample problem as consisting of three distinct densities: an urban area with a daytime population of 3000 persons/km^2 and a nighttime population of 100 persons/km^2, a suburban area with a daytime population of 200 persons/km^2 and a nighttime population of 500 persons/km^2, and a rural area with a daytime population of 10 persons/km^2 and a nighttime population of 15 persons/km^2. The route length data in Table 8-1 imply large areas and therefore large daytime total populations; the urban population is almost 200,000 people, and the suburban population also is almost 200,000 people.

In many practical situations, the population density will be found in strips parallel to the roads, particularly in suburban areas. The road width will have a characteristic density of motorists; beyond that may be a small strip of right-of-way land that will have essentially zero population; beyond that may be a few-blocks-wide strip of high-population-density stores and offices; and beyond that may be a wide area of fairly constant population density. In the case of a rail right-of-way, the no-population zone is likely larger than that for a road. It is not uncommon for either the track or the road to act as a natural barrier so that there is one land use on one side and another on the other side, with very different population densities for the two uses.

Purdy (1993) reports that a traffic buildup in the road behind an accident and a slowdown in the opposite direction contribute significantly to risk. The risk to rail passengers, at least in Great Britain, is a significant contribution to the total risk.

The effect of these variations in population density is potentially to introduce both azimuthal variation and downwind variation in the population density. Any long plume can cross from one population density zone to another (Fig. 7-6). In this example the population density values and the sizes of the corresponding areas are such that plumes are always in a single zone, and there is no azimuthal dependence. For many decisions, this is an appropriate level of detail.

For this example, the daytime population distribution will be used to evaluate the number of persons exposed. In a more detailed analysis, at least one daytime and one nighttime population distribution and their conditional probabilities might be used. The last parameter needed is the percentage of people outside. The 10% value from Section 6.5.1 is used for the urban area. The value for the suburban area is selected as 20%, and the value for the rural area is selected as 33%. The outdoor LC_{50}, which is based on 50% lethality, is 2 min, and the probability of successfully escaping indoors at this concentration and above is taken as zero. The results of combining the population density values previously given with the areas affected by the release

TABLE 8-7. Postulated lethalities for chlorine example

Scenario ID	Area (km^2)	Outdoor Lethalities (persons)			Area (km^2)	Indoor Lethalities (persons)		
		Rural	Suburban	Urban		Rural	Suburban	Urban
IGV	1.18×10^{-2}	3.89×10^{-2}	4.72×10^{-1}	3.54×10^{0}	1.48×10^{-1}	9.77×10^{-1}	2.37×10^{1}	4.00×10^{2}
RI	3.50×10^{-4}	1.16×10^{-3}	1.40×10^{-2}	1.05×10^{-1}	6.72×10^{-3}	4.44×10^{-2}	1.08×10^{0}	1.81×10^{1}
FRV (train)	3.18×10^{-3}	1.05×10^{-2}	1.27×10^{-1}	9.54×10^{-1}	3.65×10^{-2}	2.41×10^{-1}	5.84×10^{0}	9.86×10^{1}
FRV (truck)	1.32×10^{-3}	4.36×10^{-3}	5.28×10^{-2}	3.96×10^{-1}	1.85×10^{-2}	1.22×10^{-1}	2.96×10^{0}	5.00×10^{1}
IS (train)	8.55×10^{-1}	2.82×10^{0}	3.42×10^{1}	2.56×10^{2}	6.63×10^{-2}	4.38×10^{-1}	1.06×10^{1}	1.79×10^{2}
IS (truck)	3.31×10^{-1}	1.09×10^{0}	1.32×10^{1}	9.93×10^{1}	2.76×10^{-2}	1.82×10^{-1}	4.42×10^{0}	7.45×10^{1}
FWI (train)	1.26×10^{0}	4.16×10^{0}	5.04×10^{1}	3.78×10^{2}	1.14×10^{-1}	7.52×10^{-1}	1.82×10^{1}	3.08×10^{2}
FWI (truck)	4.09×10^{-1}	1.35×10^{0}	1.64×10^{1}	1.23×10^{2}	3.52×10^{-2}	2.32×10^{-1}	5.63×10^{0}	9.50×10^{1}

scenarios (Table 8-6) are shown in Table 8-7. These are the final consequence results.

8.6 RISK EVALUATION

The frequencies of accidents producing a release have been evaluated in Appendix A for 40 truck and 40 train scenarios. The results are given in Table 8-2. In the source term characterization portion of the consequence analysis (Appendix B), additional scenarios have been identified as a result of considering tank car/tank truck orientation and hole location, and at the same time similar release scenarios have been aggregated. Additional aggregation resulted in a total of eight release scenarios for detailed consequence analysis. The consequence results are given in Table 8-7. The presentation of these results in triplet, societal risk, and individual risk formats is described below.

8.6.1 Risk Triplets

Risk triplets consist of the risk scenario, the frequency, and the consequence. The triplets present some of the most useful information for determining the

TABLE 8-8. Risk triplets for truck options

Segment	Scenario	Frequency (per 180 tons delivered)	Consequence (fatalities)
Rural (US 40)	RI	2.4×10^{-5}	4.6×10^{-2}
	IGV	4.2×10^{-6}	1.0×10^{0}
	IS	5.2×10^{-6}	1.3×10^{0}
	FRV	5.9×10^{-7}	1.2×10^{-1}
	FWI	3.2×10^{-7}	1.6×10^{0}
Suburban (US 40)	RI	2.5×10^{-5}	1.1×10^{0}
	IGV	4.5×10^{-6}	2.4×10^{1}
	IS	5.4×10^{-6}	1.8×10^{1}
	FRV	6.2×10^{-7}	3.0×10^{0}
	FWI	3.4×10^{7}	2.2×10^{1}
Urban (US 40)	RI	6.3×10^{-6}	1.8×10^{1}
	IGV	1.1×10^{-6}	4.0×10^{2}
	IS	1.4×10^{-6}	1.7×10^{2}
	FRV	1.6×10^{-7}	5.0×10^{1}
	FWI	8.6×10^{-8}	2.2×10^{2}
Rural (SR 230)	RI	6.4×10^{-5}	4.6×10^{-2}
	IGV	1.2×10^{-5}	1.0×10^{0}
	IS	1.4×10^{-5}	1.3×10^{0}
	FRV	1.6×10^{-6}	1.2×10^{-1}
	FWI	8.7×10^{-7}	1.6×10^{0}

TABLE 8-9. **Risk triplets for rail option**

Segment	Scenario	Frequency (per 180 tons delivered)	Consequence (fatalities)
Rural	RI	1.6×10^{-5}	4.6×10^{-2}
	IGV	8.8×10^{-6}	1.0×10^{0}
	IS	1.0×10^{-5}	3.3×10^{0}
	FRV	4.4×10^{-6}	2.5×10^{-1}
	FWI	1.5×10^{-6}	4.9×10^{0}
Suburban	RI	3.2×10^{-6}	1.1×10^{0}
	IGV	1.8×10^{-6}	2.4×10^{1}
	IS	2.0×10^{-6}	4.5×10^{1}
	FRV	8.9×10^{-7}	6.0×10^{0}
	FWI	2.9×10^{-7}	6.9×10^{1}
Urban	RI	8.0×10^{-7}	1.8×10^{1}
	IGV	4.4×10^{-7}	4.0×10^{2}
	IS	5.0×10^{-7}	4.4×10^{2}
	FRV	2.2×10^{-7}	1.0×10^{2}
	FWI	7.2×10^{-8}	6.9×10^{2}

major risk contributors and therefore suggest the potentially most productive opportunities for reducing the risk. The frequencies shown in Table 8-2 first must be summed for each of the consequence groupings. The values in Table 8-2 are for the entire route and must be subdivided into the route segments given in Table 8-1. These frequencies are tabulated with the associated consequence from Table 8-7, and the results of this computation are given in Table 8-8 for the truck options and Table 8-9 for the train option.

8.6.2 Societal Risk

The risk triplets can be used directly to determine societal risk. They are entered into the first three columns of the scenario binning worksheet; the results for the train option are shown in Table 8-10. The results from scenario binning worksheets can be used to generate the societal risk profile in a histogram format for the train option, shown in Fig. 8-6, and for the truck options, shown in Fig. 8-7.

A more common display of societal risk is the *F-N* curve, a plot of fatalities and the frequency of *N* or more fatalities. Table 8-11 shows a worksheet used to develop the frequency of *N* or more fatalities for the train

TABLE 8-10. Scenario binning worksheet—train option

Scenario identifier	Scenario frequency	Number of people affected	Bin limits, number of people affected								
			1-2	2-5	5-10	10-20	20-50	50-100	100-200	200-500	500-1000
R-IGV	8.8×10^{-6}	1	8.8×10^{-6}								
R-IS	1.0×10^{-5}	3.3		1.0×10^{-5}							
R-FWI	1.5×10^{-6}	4.9		1.5×10^{-6}							
S-RI	3.2×10^{-6}	1.1	3.2×10^{-6}								
S-IGV	1.8×10^{-6}	24					1.8×10^{-6}				
S-IS	2.0×10^{-6}	45					2.0×10^{-6}				
S-FRV	8.9×10^{-7}	6			8.9×10^{-7}						
S-FWI	2.9×10^{-7}	69						2.9×10^{-7}			
U-RI	8.0×10^{-7}	18				8.0×10^{-7}					
U-IGV	4.4×10^{-7}	400								4.4×10^{-7}	
U-IS	5.0×10^{-7}	440								5.0×10^{-7}	
U-FRV	2.2×10^{-7}	100							2.2×10^{-7}		
U-FWI	7.2×10^{-8}	690									7.2×10^{-8}
Total frequency			1.2×10^{-5}	1.2×10^{-5}	8.9×10^{-7}	8.0×10^{-7}	3.8×10^{-6}	2.9×10^{-7}	2.2×10^{-7}	9.1×10^{-7}	7.2×10^{-8}

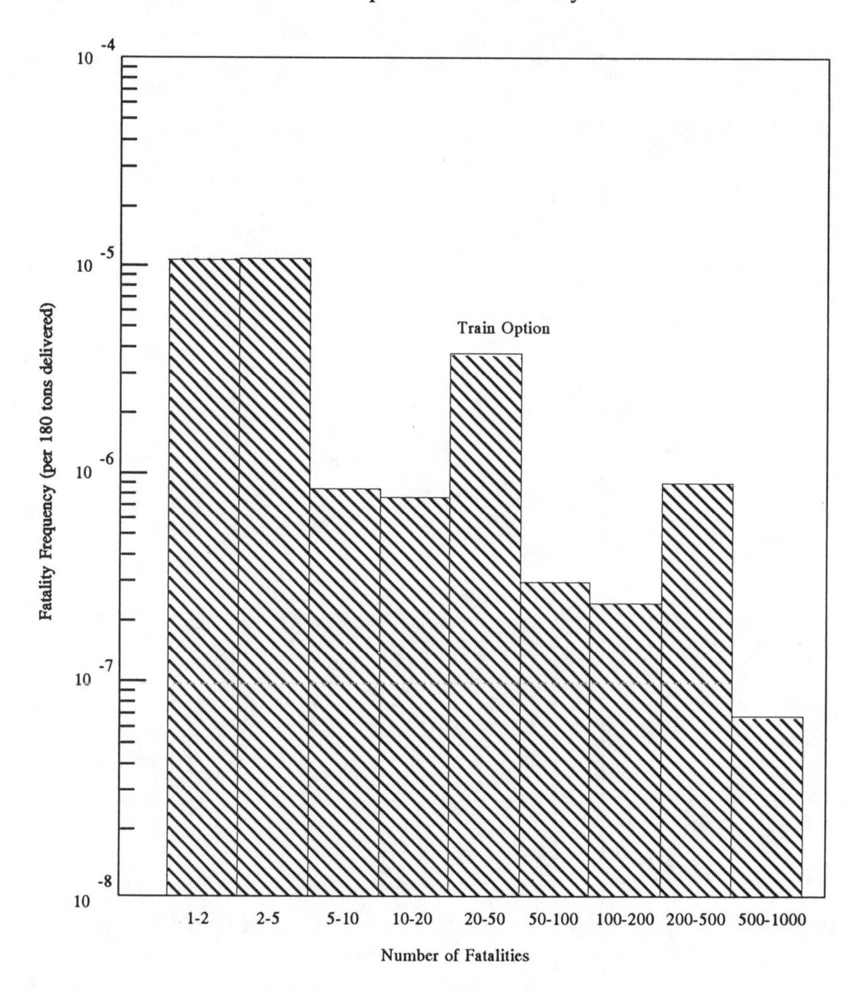

FIGURE 8-6. Societal risk histogram for train option.

option from the data in Table 8-10. Figure 8-8 is an *F-N* plot for all three truck and train options in this example.

8.6.3 Individual Risk

The first two data elements of the risk triplets can be used to determine individual risk profiles. The individual risk worksheets for the train option are shown in Tables 8-12 and 8-13, and the individual risk profiles for all three options are shown in Fig. 8-9.

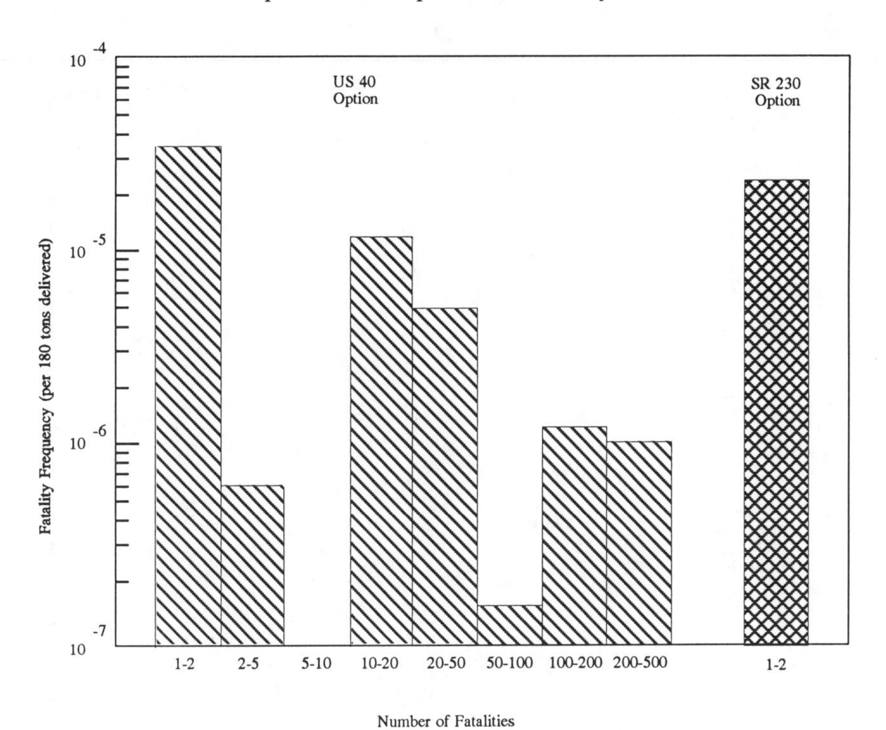

FIGURE 8-7. Societal risk histogram for truck options.

TABLE 8-11. Cumulative frequency worksheet—train option

Bin limit	Frequency	Cumulative frequency
1-2	1.2×10^{-5}	3.1×10^{-5}
2-5	1.2×10^{-5}	1.9×10^{-5}
5-10	8.9×10^{-7}	7.0×10^{-6}
10-20	8.0×10^{-7}	6.0×10^{-6}
20-50	3.8×10^{-6}	5.3×10^{-6}
50-100	2.9×10^{-7}	1.5×10^{-6}
100-200	2.2×10^{-7}	1.2×10^{-6}
200-500	9.1×10^{-7}	9.8×10^{-7}
500-1000	7.2×10^{-8}	7.2×10^{-8}

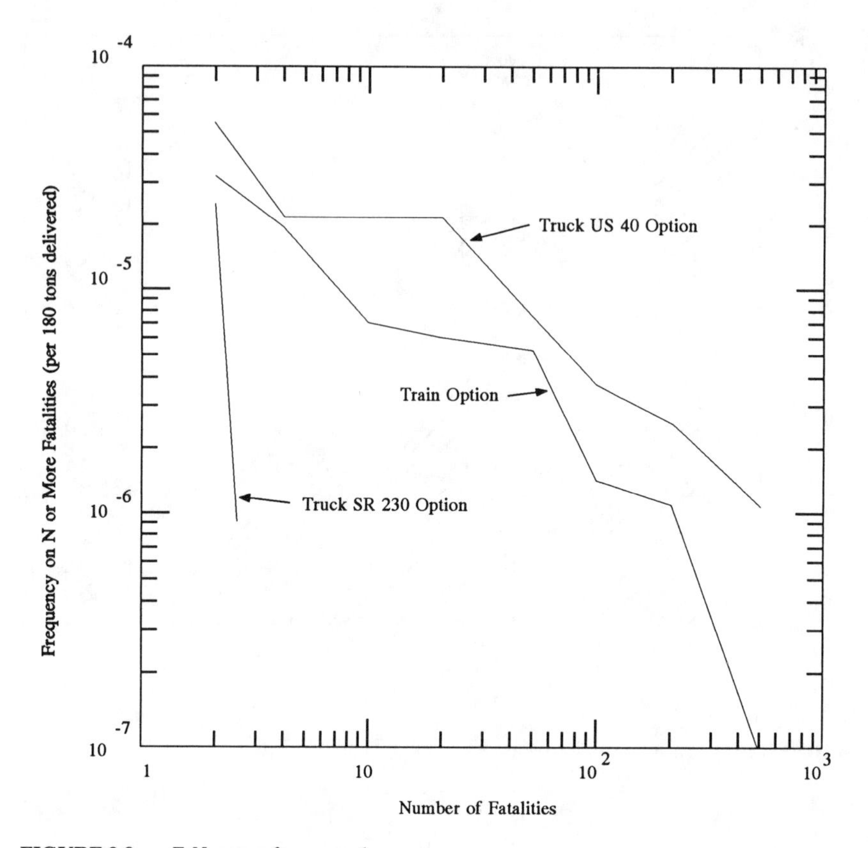

FIGURE 8-8. *F-N* curves for example.

TABLE 8-12. Individual risk calculation worksheet no. 1

Case Identification: Train
Route Segment Identification: Urban (All will be equal when segment length corrected.)
Comments: The larger of the outdoor or indoor lethal distance is used. Frequency units are per 180 tons delivered.

Scenario identification	Release frequency	Arc probability	Total frequency	Lethal distance, m
RI	8.00×10^{-7}	0.5	4.00×10^{-7}	330
IGV	4.04×10^{-7}	0.5	2.00×10^{-7}	1550
IS	4.68×10^{-7}	0.5	2.34×10^{-7}	2100
FRV	2.22×10^{-7}	0.5	1.11×10^{-7}	820
FWI	7.24×10^{-8}	0.5	3.62×10^{-8}	2500

TABLE 8-13. Individual risk calculation worksheet no. 2

Case Identification: Train
Route Segment Identification: Urban (All will be equal after the chord correction.)
Comments: Frequency units are per 180 tons delivered. Urban segment length is 5 miles or 8045 meters.

		Lethal Distance, m									
		330		820		1550		2100		2500	
Scenario identification	Scenario frequency	Chord correction	Modified frequency	Chord correction	Modified frequency	Chord correction	Modified frequency	Chord correction	Modified frequency	Chord correction	Modified frequency
FWI	3.62×10^{-8}	6.16×10^{-1}	2.23×10^{-8}	5.87×10^{-1}	2.12×10^{-8}	4.88×10^{-1}	1.77×10^{-8}	3.37×10^{-1}	1.22×10^{-8}	1.55×10^{-2}	5.61×10^{-10}
IS	2.34×10^{-7}	5.16×10^{-1}	1.21×10^{-7}	4.81×10^{-1}	1.12×10^{-7}	3.52×10^{-1}	8.24×10^{-8}	1.30×10^{-2}	3.04×10^{-9}		
IGV	2.02×10^{-7}	3.76×10^{-1}	7.60×10^{-8}	3.27×10^{-1}	6.60×10^{-8}	9.63×10^{-3}	1.94×10^{-9}				
FRV	1.11×10^{-7}	1.87×10^{-1}	2.08×10^{-8}	5.10×10^{-3}	5.66×10^{-10}						
RI	4.00×10^{-7}	2.05×10^{-3}	8.20×10^{-10}								
Total			2.48×10^{-7}		2.00×10^{-7}		1.02×10^{-7}		1.52×10^{-8}		5.61×10^{-10}

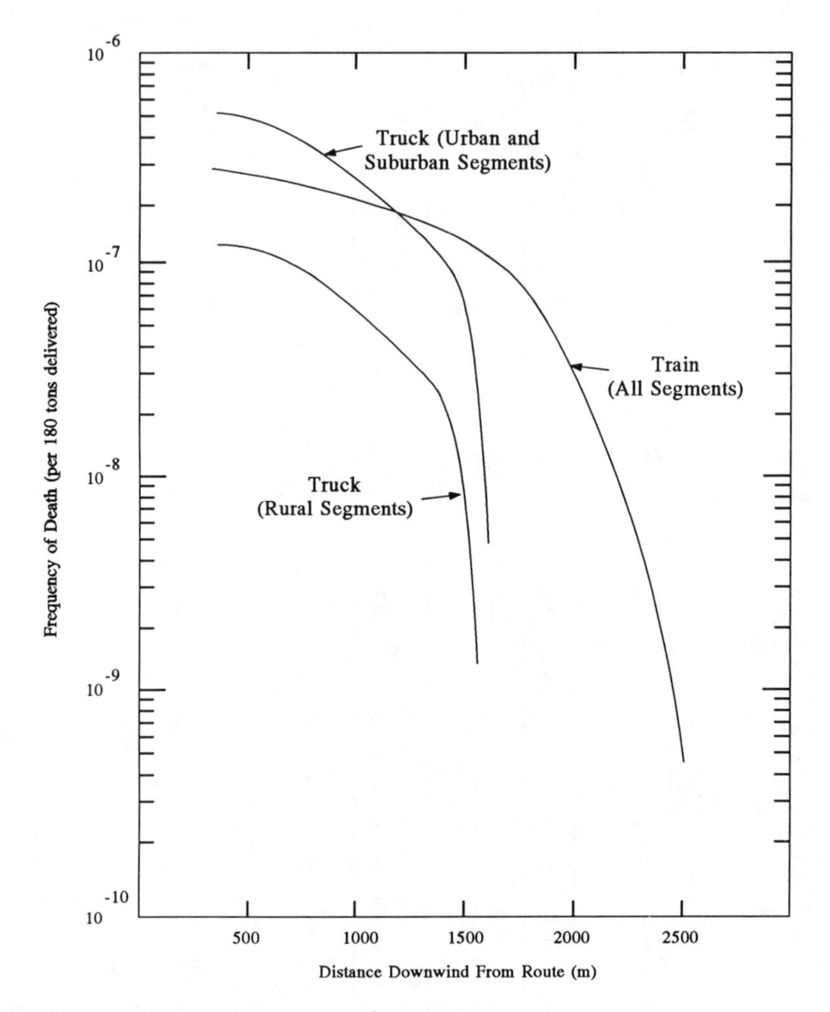

FIGURE 8-9. Individual risk contours for example.

8.7 RISK REDUCTION ALTERNATIVES

The objective of the analysis described in this chapter is to determine the best risk reduction alternative from the three choices described in Section 8.1. The results presented in Figs. 8-6 through 8-9 show that the truck option utilizing the SR 230 route results in the lowest societal and individual risk. This book will not address the risk acceptability issue (risk assessment); however, almost any risk analysis will indicate ways that risk can be reduced (risk management). The risk analyst always should indicate to the decision maker those risk reduction measures that might be cost-effective. Two major contributors to this risk analysis are the nonaccident relief valve actuation for the truck options and the impact failure of the relief valve for the truck and train options. These two contributors and suggestions for reducing their effect are now described.

A major contributor to the truck risk is the nonaccident relief valve actuation scenario, indicated as RI. Data based on all commodity releases by truck showed that a nonaccident release from the safety valve was 2.3 times more likely than an accident release. Thus, the RI scenario is by definition the single largest contributor to release frequency. The consequence model for the relief valve actuation is based on a plume of several hours' duration. People outdoors will easily escape the narrow plume with little chance of a fatality, but people indoors may be trapped for some time; thus a relatively small concentration over a relatively long duration is calculated to cause few fatalities in the rural area, 1 to 2 fatalities in the suburban area, and 10 to 20 fatalities in the urban area. Improvements in valve maintenance and post-loading valve securement can reduce the frequency of this moderately severe consequence scenario.

The impact fails valve scenario, or IGV, exhibits the same long-plume behavior as the RI scenario and is the second largest frequency event for truck transport and third largest for train transport. The IGV consequences contribute the largest single number of fatalities for the truck suburban (24 fatalities) and urban (400 fatalities) segments and the third largest for the train consequences (400 fatalities for the urban segment). The likelihood of rescue or action to stop or reduce the release from either the safety valve (RI) or the impact damaged gas valve (IGV) can be improved by the training of fire and other emergency response personnel to quickly and correctly respond to chlorine releases. Both the truck and the train consequence component of risk will be reduced by this training, but the scenario frequency will be unchanged.

Risk reduction measures, such as improved driver training, improved inspection and maintenance, and changes in container design, can be explicitly evaluated for risk reduction potential. The evaluation may be qualitative or quantitative; both the effectiveness and the cost need to be evaluated. These additional risk reduction alternatives are usually evaluated only for the most viable route and mode alternative, in this case the highway route through the low-population area. Care must be exercised, however, that the original route or mode selection is not compromised by the additional risk reduction measures.

A major feature of the bottom-up, or predictive, approach to transportation QRA is the ability to quantitatively evaluate the risk reduction benefits of overspecifying the transport container. The DOT specifications define the minimum allowable container, and it is permissible to use a container that exceeds the minimum, that is, to overspecify. Failure by impact (IS) is a major frequency and consequence contributor to both truck and train risk, as can be seen in Tables 8-8 and 8-9. The impact failure threshold, as well as crush, puncture, and fire thresholds, for the overspecified container would be required for the analysis. The frequency analysis would be reevaluated to determine the reduction in frequency achieved. Some aspects of the consequence portion of the calculation may require reevaluation. The net reduction in risk can be evaluated quantitatively to determine whether the proposed change is cost-effective.

The bottom-up approach also can be used to determine design parameters needed to meet a particular frequency or a particular risk goal. The procedure is to select a set of parameters and evaluate the frequency or the risk as appropriate. Depending on whether the target is undershot or overshot, parameter(s) are revised and the calculations repeated until the calculated risk is sufficiently close to the risk goal.

References

Kirchner, J. R., and W. R. Rhyne. 1992. The risks of handling vs. transporting dangerous goods. Paper read at the International Consensus Conference: Risks of Transporting Dangerous Goods, April 6-8, University of Waterloo, Toronto, Canada.

Marshall, V. C. 1987. *Major Chemical Hazards.* Chichester, England: Ellis Horwood Limited.

Purdy, G. 1993. Risk analysis of the transportation of dangerous goods by road and rail. *Journal of Hazardous Materials* 33(2): 229-259.

Trinity Consultants, Inc. 1991. *User's Manual for DEGADIS+: Version 2.0.* Dallas, Texas: Trinity Consultants, Inc.

9

Summary

This chapter summarizes the important features of transportation QRA, emphasizing those features that distinguish it from stationary chemical or nuclear facility QRA.

9.1 REASONS FOR THE ANALYSIS

The U.S. Department of Transportation (DOT) has set standards for containers that are used for transporting hazardous materials; therefore, a potential shipper or transporter may question the need for a risk analysis if the shipment is in compliance with DOT regulations. Four reasons to perform a QRA are given in Section 2.1, where the underlying theme is that a QRA provides a basis for making consistent and defendable decisions. The most important decision is to determine how to cost-effectively reduce risks with alternative procedures or designs.

9.2 COMPARISON WITH CHEMICAL PROCESS QRA

The methodologies for transportation and for chemical process QRAs are basically the same (Fig. 1.1). The major differences are in the databases used for determining the frequency of an accident producing a release and in the requirement to repeat the analysis at a number of potential accident sites along the route(s).

If a truck is in an accident, a number of things can happen. For example, the truck can impact a hard (bridge abutment) or a soft (motorcycle) object, and it can overturn or stay upright after the impact. The data that describe the potential truck or train accident scenarios leading to a release of hazardous material are limited. The approach used to analyze transportation scenarios is to simplify the accident forces: perhaps to fire, puncture, impact, and crush, or perhaps to just one force.

Once accident forces have been identified, the standard QRA procedure is to use fault trees and/or event trees to develop accident scenarios. A unique feature of transportation QRA is that fault trees and event trees may not have to be generated essentially from scratch. Because containers on wheels are similar and experience similar accident environments, it usually is feasible to start with a standard fault tree or a generalized event tree and modify it as needed. Transportation event trees and fault trees are presented in Chapter 4. Fault trees and event trees can be constructed to many levels of detail, depending on the decision for which the analysis is being made.

In a specific transportation accident scenario, the response of the container to mechanical forces or to a thermal force is modeled first. Chapter 5 contains a description of container failure analysis models. Given a rupture in the transport container, a dynamic model, as described in Chapter 6, may be used to calculate a release rate. In a chemical process facility, a dynamic analysis of a system of pipes, pumps, and so forth may be needed first to determine the flow rates, pressures, and so on that threaten the containment of the hazardous material. Given a failure, the analytical model then used to estimate release rates probably is more complicated for a process facility than for the transport container. An exception is simple storage tanks at the stationary facility.

Consequence models are similar for both QRA types, and the consequence analysis can have a significant effect on risk analysis uncertainty. The selection of meteorological parameters can have orders of magnitude effects on the number of persons calculated to be exposed to a release. The estimation of health effects from a calculated exposure can be subject to a great deal of uncertainty due to lack of data. These potentially weak links in the analysis can affect decisions about the other analysis steps. If a great deal of uncertainty is associated with one analytical step, a very detailed, and costly, analysis in another analytical step is questionable; see Section 2.7.

A major difference between transportation and stationary plant QRAs is that the transportation consequence analysis must be performed at multiple locations along the route or routes. The computational scheme can be set up so that the cost of multiple frequency analyses is relatively small; however, multiple-site consequence analysis can be difficult to perform without significant cost escalation. The example presented in Chapter 8 kept the cost down by assuming that the same meteorology applied at all analysis locations, population distributions were uniform at each accident location, and there was no azimuthal parameter variation at any accident location.

9.3 APPROACHES FOR TRANSPORTATION QRA

The two basic approaches for transportation QRA differ in the way that the frequency analysis is performed. Each approach uses data to estimate the frequency of accident releases and the amount of the release. The difference is illustrated by the action or actions taken when directly applicable data are not available. In the top-down, or historical, approach, the definition of the hazardous material is broadened to include materials of similar behavior, and/or the container definition is broadened to include similar containers until enough historical data are included in the definition to permit derivation of a statistically significant value for the parameter. In the bottom-up, or predictive, approach, the analyst uses engineering models and the available data to construct a prediction for the needed parameter. No one approach is best for general use; all QRAs aggregate to some extent. The risk analyst must decide how to approach each problem. This book focuses on the more computationally complex predictive, or bottom-up, approach; see Sections 2.4 and 2.5.

9.4 THE MATHEMATICAL FORMULATION

The risk, R_i, for accident scenario i is a function of the scenario frequency, F_i, and the scenario consequence, C_i:

$$R_i = f(F_i, C_i) \tag{9-1}$$

The overall risk is obtained by summing over all scenarios:

$$R = \Sigma R_i \tag{9-2}$$

The scenario frequency computation usually is divided into three components: the accident frequency; the conditional probability of a release, given an accident; and the conditional probability of various consequence terms. The accident frequency starts with a value for accidents per mile and usually ends with accidents per year or accidents per some unit of material delivered so that all analyses can be put on a common basis. The conditional probability of release may be subdivided into several components in the predictive approach or simply evaluated at this top level in the historical approach. The consequence analysis usually introduces some conditional probabilities into the frequency term, such as the probability that a certain meteorological condition exists, given that the accident has occurred. The terms in the mathematical formulation vary with the specific analysis; Eqs. 2-2 and 7-1 are typical.

9.5 THE TRANSPORTATION DATABASE

The accident frequency is computed by dividing the number of accidents by the corresponding exposure, that is, the number of miles. The number of fatal accidents is reasonably reliably recorded for trucks, but the nonfatal accidents are not. The truck exposure data suffer from a lack of consistency among data sources. In addition, the accident data do not correspond well with the exposure data. Accidents are based on police reports, and the mileage is based on questionnaires or traffic counts. The choices available to the risk analyst for obtaining truck accident frequency are either (1) to obtain databases and compute values or (2) to use one of the existing database analyses. The use of available analyses is cost-effective and adequate for most decisions. The uncertainty in the reported analyses is small compared to other uncertainties in both the frequency and the consequence components.

Train accident frequency analysis is more straightforward than that for trucks because there is essentially one source of data for both the number of accidents and the number of train miles. Accident characteristics are much better documented for trains than for trucks. A notable train data deficiency is the lack of an accident frequency as a function of track class.

Accident force types and force magnitudes for predictive analyses were developed in the 1970s. It is possible to verify the continuing appropriateness of much of the input data, but it is difficult to verify the final results without complete redevelopment and reanalysis of the engineering models. Four useful accident environments were defined: impact, crush, puncture, and fire. (The remaining force, immersion, is not generally applicable.) Analytical models, such as a spring-mass description of the tractor, trailer, and cargo when striking a hard surface, were used to generate cumulative probability distributions for the magnitude of the accident forces. The cumulative probability distribution results have a large uncertainty, about one order of magnitude, for a specific container being analyzed. These models and probability distributions are described in Section 3.2.

The accident force required to fail a container is computed by using techniques described in Chapter 5. The cumulative probability results of Section 3.2 are used to determine the conditional probability of failure, given an accident. The probability of failure using this approach can explicitly incorporate container design features in varying levels of detail. The failure threshold uncertainty can be reduced, but not eliminated, by more sophisticated and expensive calculations.

Conditional container failure probability data are presented in Section 3.3. Truck data are available when aggregated by either failure type, cargo type, accident type, or highway type. Train data for tank cars are available for all hazardous materials aggregated together, or by tank car group and hazardous material class. For a particular accident involving a specific hazardous material in a specific container, the uncertainty is in an order of magnitude range. Using data from Section 3.3 is quick, but the ability to use container design as a risk mitigation tool is essentially nil.

Few published data exist for the amount of hazardous material released, given a release. The Hazardous Material Incident System is a great potential source of raw data for a systematic evaluation of the needed parameter(s) for quantitative risk analysis.

9.6 RESULTS PRESENTATION

The standard results presentation techniques are used in transportation QRA: risk triplets, societal risk, and individual risk. Risk triplets are too detailed for decision making but are useful to the risk analyst for proposing cost-effective risk mitigation measures. Societal risk is the usual risk presentation technique although the individual risk technique is useful in some circumstances. The example calculation presented in Chapter 8 develops all three risk presentation techniques in detail to improve the reader's understanding of these concepts.

Appendix A

Numerical Evaluation of Train and Truck Accident Scenario Frequencies

The calculation of the frequency values presented in Table 8-2 is described in this appendix. The train evaluation is presented first, followed by the truck evaluation for the SR 230 option. The frequencies for truck transport via the US 40 option are obtained from the SR 230 values by the ratio of accident frequencies. The way that the data are presented in Chapter 3 will occasionally make one-for-one application difficult in the fault tree presented in Chapter 4.

A.1 TRAIN ACCIDENT SCENARIOS

Train Impact Accidents

The basic event "impact occurs on tank head" in Fig. 4-5 includes three factors: the frequency that an accident occurs, the conditional probability that an impact force results from the accident, and the conditional probability that the impact force is applied to the tank head. The accident frequency was determined in Figure 8-3 to be 2.58×10^{-3} accidents/180 tons delivered. Impact forces arise from either a collision or a derailment accident; the conditional probability of either is 0.90, given a train accident (Section 3.2.3.2). Impact is assumed to occur on the tank head if the initial accident is a head-on, rear-end, broken-train, or obstruction collision. These accidents constitute 22% of the collision accidents in Table 3-24. It is assumed that 70% of derailments involve head impacts. Train accidents are 13% collisions and 77% derailments without collisions (Fig. 3-11), and the conditional probability of a head impact, given an impact accident is:

$$[(0.22)(0.13) + (0.70)(0.77)]/(0.13 + 0.77) = 0.630$$

The basic event "impact occurs on tank head" is evaluated as:

$$(2.58 \times 10^{-3} \text{ train accidents/180 tons})(0.90 \text{ impact accidents/train accident})$$
$$(0.630 \text{ head impacts/impact accident}) = 1.46 \times 10^{-3} \text{ head impacts/180 tons}$$

The conditional probability of impact on the manway and the valve enclosure usually is estimated on the basis of area ratios. If the object impacted is small (e.g., a bridge column), the ratio of manway area to tank area, about 0.003, appears to be the appropriate probability. If the object is large (e.g., an embankment), the percent of the circumference that the manway covers, 10%, appears to be a better estimate. The rollover type of derailment (Fig. 3-11) frequently involves sliding of the tank car on its side. Given this orientation, the probability of subsequent impact on the valve cover would seem to be higher than 0.10, say 0.25. It is assumed that each of the three values 0.003, 0.10, and 0.25 is equally probable, for an average of 1.18×10^{-1}. This value is the conditional probability of manway or valve-cover impact, given a shell impact. The conditional probability of manway or valve cover impact, given any impact, is 4.4×10^{-2}. The frequency of a manway or valve-cover impact is evaluated as:

$$(2.58 \times 10^{-3} \text{ train accidents/180 tons})(0.90 \text{ impact accidents/train accident})$$
$$(4.4 \times 10^{-2} \text{ manway impacts/impact accident})$$
$$= 1.02 \times 10^{-4} \text{ manway impacts/180 tons}$$

The frequency of shell impact is evaluated similarly as:

$$(2.58 \times 10^{-3} \text{ train accidents/180 tons})(0.90 \text{ impact accidents/train accident})$$
$$(0.326 \text{ shell impacts/impact accident}) = 7.57 \times 10^{-4} \text{ shell impacts/180 tons}$$

The probability that an excess flow valve fails to close upon demand is 2.2×10^{-3} (CCPS 1989). The probability that the tank shell, the tank head, or the manway is defective is estimated as 2.6×10^{-3} (Southwest Research Institute 1979). This value is based on four failures from material defects in 1552 pressurized liquid tanks. The effect of wall thinning with age usually is not considered as a defect in this context. Rather, the failure threshold calculation (Chapter 5) usually includes manufacturing tolerances as well as corrosion and other operational factors.

The probability that the impact force magnitude is large enough to cause failure that leads to a release is evaluated from Table 3-15, which gives the cumulative frequency that the accident velocity change is less than the tabulated value. For risk analyses, the parameter of interest is the cumulative frequency of exceeding the failure threshold, or 1 minus the value of being less than the threshold velocity value. The cumulative frequency values in Table 3-15 are per train accident; therefore, these values must be multiplied by 0.90 impact accidents/train accident to obtain units compatible with the calculations for this example. To use Table 3-15, it is necessary to assume that only one chlorine tank car is included in each shipment. The conditional probabilities of the impact failure, given an impact, for this example, based on the failure thresholds in Table 5-1, are given in Table A-1.

TABLE A-1. Probabilities of train example impact failures

Parameter	Probability given an impact accident	Basis for evaluation
Impact fails tank head	2.07×10^{-3}	The head fails at 32 mph from impact (Table 5-1). Cases 1 and 2 in Table 3-15 are averaged to obtain 0.99770 for 32 mph. The probability of exceeding this impact velocity is $1 - 0.99770 = 0.0023$ per train accident.
Impact fails tank shell	8.82×10^{-3}	The shell fails at 18 mph from impact (Table 5-1). Cases 1 and 2 in Table 3-15 are averaged to obtain 0.99020 for 18 mph. The probability of exceeding this impact velocity is $1 - 0.99020 = 0.00980$ per train accident.
Impact fails defective tank shell	2.38×10^{-2}	Andrews et al. (1980) calculate a failure threshold of 9 mph also for a defective tank shell.
Impact fails defective manway	2.38×10^{-2}	A value of 9 mph is assumed to be the failure threshold for a defective manway.
Impact fails liquid or gas valves	1.06×10^{-1}	Given that impact occurs on the valve dome, any nonzero velocity is assumed to fail the valves. Cases 1 and 2 in Table 3-15 are averaged to obtain 0.88174. $1 - 0.88174 = 0.11826$ per train accident.
Impact fails manway	8.82×10^{-3}	The failure of interest is the connection between the tank shell and the manway cover. This area is the same thickness as the shell; therefore, the 18-mph value for the tank shell failure is used as the failure value.
Impact fails defective tank head	2.38×10^{-2}	Andrews et al. (1980) calculate a failure threshold of 9 mph for a defective tank head. Cases 1 and 2 in Table 3-15 are averaged to obtain 0.97353. The probability of exceeding this value is $1 - 0.97353 = 0.02647$ per train accident.

Enough information is now available to evaluate the impact scenarios listed in Fig. 8-2 and illustrated in Fig. 4-5.

ILV—Impact fails liquid valve and excess flow valve does not close
Impact occurs on valve dome = 1.02×10^{-4} valve impacts/180 tons transported
Impact sufficient to fail liquid valve = 1.06×10^{-1} failures/impact
Excess flow valve fails to close = 2.2×10^{-3} failures/demand
ILV frequency = 2.4×10^{-8} failures/180 tons delivered

IGV—Impact fails gas valve
Impact occurs on valve dome = 1.02×10^{-4} valve impacts/180 tons transported
Impact sufficient to fail gas valve = 1.06×10^{-1} failures/impact
IGV frequency = 1.1×10^{-5} failures/180 tons delivered

ISD—Impact fails defective tank shell
Impact occurs on tank shell = 7.57×10^{-4} shell impacts/180 tons transported
Shell is defective = 2.6×10^{-3} defective shell/shell
Impact sufficient to fail defective shell = 2.38×10^{-2} failures/shell impact
ISD frequency = 4.7×10^{-8} failures/180 tons delivered

IS—Impact fails tank shell
Impact occurs on tank shell = 7.57×10^{-4} shell impacts/180 tons transported
Impact sufficient to fail tank shell = 8.82×10^{-3} failures/shell impact
IS frequency = 6.7×10^{-6} failures/180 tons delivered

IHD—Impact fails defective tank head
Impact occurs on tank head = 1.46×10^{-3} head impacts/180 tons transported
Head is defective = 2.6×10^{-3} defective head/head
Impact sufficient to fail defective head = 2.38×10^{-2} failures/impact
IHD frequency = 9.0×10^{-8} failures/180 tons delivered

IH—Impact fails tank head
Impact occurs on tank head = 1.46×10^{-3} head impacts/180 tons transported
Impact sufficient to fail head = 2.07×10^{-3} failures/impact
IH frequency = 3.0×10^{-6} failures/180 tons delivered

IMD—Impact fails defective manway
Impact occurs on manway = 1.02×10^{-4} manway impacts/180 tons transported
Manway is defective = 2.6×10^{-3} defective manway/manway
Impact sufficient to fail defective manway = 2.38×10^{-2} failures/impact
IMD frequency = 6.3×10^{-9} failures/180 tons delivered

IM—Impact fails manway
Impact occurs on manway = 1.02×10^{-4} manway impacts/180 tons transported
Impact sufficient to fail manway = 8.82×10^{-3} failures/impact
IM frequency = 9.0×10^{-7} failures/180 tons delivered

The probability that a train accident producing impact forces, that is, a collision or derailment accident, produces fire is 0.01 (Section 3.2.3.1). The frequencies of the impact scenarios and the frequencies of the impact scenarios followed by fire now can be entered into Table 8-2.

Train Puncture Accidents

The tank shell thickness is 0.787 in., and the equivalent head thickness is 1.16 in. (Table 5-1). The probability of a puncture for these dimensions is obtained from Table 3-16 as 5.71×10^{-4} punctures/accident for the shell thickness and 4.44×10^{-4} punctures/accident for the head thickness. The valve dome thickness is 0.375 in., but the minimum thickness value in Table 3-16 of 0.4375 in. is used, and the result is 7.41×10^{-4} punctures/accident. These units mean that the conditional probability that a puncture situation occurs, given an accident, has been included in the values obtained from Table 3-16.

Impact and puncture forces both arise from collision and derailment accidents; therefore, the distribution derived earlier for impact location is assumed to be applicable for puncture location with the exception that the area ratio from the impact analysis, 0.003, is the appropriate conditional probability for valve dome puncture, given shell puncture. Thus, the conditional probability of a probe contacting the tank head is 0.630, the conditional probability of a probe contacting the tank shell is 0.369, and the conditional probability of a probe contacting the valve dome is 0.001.

If a probe punctures the valve dome, it is assumed, on the basis of the number of valves, to fail a liquid valve 40% of the time, a gas valve 40% of the time, and a safety valve 20% of the time. The assumption that each valve dome puncture will result in the failure of one valve will tend to offset the lower computed valve cover penetration probability due to the limitations of Table 3-16. The probability of multiple valve punctures is assumed to be negligible.

The use of values from Table 3-16 results in units for the "probe impact sufficient to penetrate head" event that includes aspects of the "puncture situation occurs" event as the two are defined in Fig. 4-6. The accident frequency is contained in the units from Table 3-16; whereas, as defined in Fig. 4-6, the accident frequency is included in the "puncture situation occurs" event. The accident frequency can be included in either place as long as it is properly accounted for. The use of explicit units on the numerical values, as is done in the calculations of this section, is recommended to prevent mistakes. The scenario frequencies now can be calculated.

PH—Puncture probe fails tank head
Train accident occurs = 2.58×10^{-3} train accidents/180 tons delivered
Puncture probe contacts tank head = 0.630 head punctures/puncture
Puncture failure occurs = 4.44×10^{-4} punctures/accident
PH frequency = 7.2×10^{-7} punctures/180 tons delivered

PS—Puncture probe fails tank shell
Train accident occurs = 2.58×10^{-3} train accidents/180 tons delivered
Puncture probe contacts tank shell = 0.369 shell punctures/puncture
Puncture failure occurs = 5.71×10^{-4} punctures/accident
PS frequency = 5.4×10^{-7} punctures/180 tons delivered

PV—Puncture probe fails valve dome
Train accident occurs = 2.58×10^{-3} train accidents/180 tons delivered
Puncture probe contacts valve dome = 1.1×10^{-3} valve dome punctures/puncture
Puncture failure occurs = 7.41×10^{-4} valve dome punctures/accident
PV frequency = 2.1×10^{-9} punctures/180 tons delivered

The probability of fire following puncture is 0.01 (Section 3.2.3.1), and the frequencies of scenarios involving puncture and puncture followed by fire now can be entered into Table 8-2.

Train Crush Accidents

The conditional probability of a crush force is 2×10^{-3} per train accident (Section 3.2.3.3). The frequency corresponding to crush forces applied to a tank car is evaluated as (2.58×10^{-3} train accidents/180 tons)(2×10^{-3} crush accidents/train accident) = 5.16×10^{-6} crush accidents/180 tons delivered. For the 134,000-lb crush failure threshold given in Table 5-1, the failure probability is evaluated from Fig. 3-13 as 1.0 − 0.92 = 0.08. The failure threshold for a defective shell is assumed to be 75% of this value, 100,500 lb, and the failure probability is evaluated as 1.0 − 0.67 = 0.33.

Enough information is now available to evaluate the crush scenarios listed in Table 8-2 and illustrated in Fig. 4-7.

CD—Crush fails defective tank
Crush forces applied to tank = 5.16×10^{-6} crush accidents/180 tons transported
Tank is defective = 2.6×10^{-3} defective tank/tank
Crush sufficient to fail tank = 0.33 failures/crush accident
CD frequency = 4.4×10^{-9} failures/180 tons transported

C—Crush fails tank
Crush forces applied to tank = 5.16×10^{-6} crush accidents/180 tons transported
Crush sufficient to fail tank = 0.08 failures/crush accident
C frequency = 4.1×10^{-7} failures/180 tons transported

The probability of fire following crush, that is, a derailment accident, is 0.01 (Section 3.2.3.1), and the frequencies of crush and crush followed by fire now can be entered into Table 8-2.

Train Fire Accidents

The fault tree in Fig. 4-8 shows the potential scenarios for container failure due to accidents involving only fire forces. The probability of fire, given a train accident, is 0.184 (Section 3.2.3.1); however, this value does not reflect whether the chlorine tank car is involved in the fire. Assuming that the trains delivering the chlorine in the example have a total of 50 railcars, a chlorine tank car involvement of 2.02×10^{-2}/train fire is obtained by using the correction factor suggested in Section 3.2.3.1. The "fire occurs" basic event is then:

(2.58×10^{-3} train accidents/180 tons)(2.02×10^{-2} fires of interest/train fire)
= 5.21×10^{-5} fires of interest/180 tons

The fault tree in Fig. 4-9 shows the potential scenarios for an accident producing mechanical damage (but no failure) followed by a fire that causes the weakened container to fail. The additional value needed for this analysis is the probability that an impact force is sufficient to weaken, but not fail, the container. Impacts between the value required to fail a defective shell (9 mph) and the value required to fail a nondefective shell (18 mph) are assumed to substantially weaken the shell. This level of impact is assumed also to cause a 10% insulation loss. The conditional probability of a weakened container with insulation damage, given an impact accident, is the difference between the 9- and 18-mph values in Table A-1, or 1.50×10^{-2}. The conditional probability of insulation damage but no degradation in tank strength is assumed to be an impact accident at less than 9 mph. The probability of this accident, which is the probability of a 9-mph impact or less, is 0.97353 minus the probability of no impact, which is 0.88174 (Table A-1). The probability of an impact, but an impact less than 9 mph, is then 0.0918 per train accident.

The probability that a relief valve fails to open upon demand is 2.1×10^{-4} (CCPS 1989). The probability that a car involved in fire is upright is estimated in Section 3.2.3.1 as 0.69. The corresponding value for an overturned car is 0.31. The probabilities of the train fire failures for this example, based on the fire failure threshold in Table 5-1, are given in Table A-2. These values

TABLE A-2. Probabilities of train example fire failures

Parameter	Probability of failure	Basis for evaluation
Fire duration raises pressure to relief valve pressure, 100 min	0.12	The probability of exceeding 100 min is evaluated from Fig. 3-10 as $1 - 0.88 = 0.12$.
Fire duration fails weakened or defective shell, 100 min	0.12	The container is assumed to rupture at the relief valve start-to-discharge pressure of 2.69×10^6 N/m^2 (375 psig).
Fire duration ejects > 50% of inventory, 290 min	0.01	The maximum fire duration indicated in Fig. 3-10 is 160 min. The probability of exceeding 290 min is estimated as $1 - 0.99 = 0.01$.
Fire duration lifts relief valve but does not result in ejection of > 50% of inventory	0.11	The probability of a fire that exceeds 100 min (relief valve lifts) but does not exceed 290 min (> 50% inventory loss) is $0.12 - 0.01 = 0.11$.
Fire duration causes overpressure with liquid discharge, 164 min	0.06	The maximum fire duration indicated in Fig. 3-10 is 160 min, which will be used to evaluate this parameter. The probability of exceeding 160 min is evaluated as $1 - 0.94 = 0.06$.
Fire duration lifts relief valve but does not result in overpressure; discharge is liquid only	0.06	The probability of a fire that exceeds 100 min (relief valve lifts) but does not exceed 164 min (overpressurization) is $0.12 - 0.06 = 0.06$.
Weakened or defective tank with insulation damage fails in fire of 35 min duration	0.41	The probability of exceeding 35 min (relief valve lifts) is evaluated from Fig. 3-10 as $1 - 0.59 = 0.41$.
Tank with insulation damage ejects > 50% of inventory due to fire of 100 min	0.12	The probability of exceeding 100 min is evaluated from Fig. 3-10 as $1 - 0.88 = 0.12$.
Relief valve of tank with insulation damage lifts from fire, but ejection of > 50% of inventory does not occur	0.29	The probability of a fire that exceeds 35 min (relief valve lifts) but not exceeding 100 min (> 50% inventory loss) is $0.41 - 0.12 = 0.29$.
Tank with insulation damage and liquid discharge fails in fire due to overpressure, 55 min	0.18	The probability of exceeding 55 min is evaluated from Fig. 3-10 as $1 - 0.82 = 0.18$.
Relief valve of tank with insulation damage lifts from fire, but overpressurization does not occur with only liquid discharge	0.23	The probability of a fire exceeding 35 min (relief valve lifts) but not exceeding 55 min (overpressurization) is $0.41 - 0.18 = 0.23$.
Relief valve of tank with insulation damage lifts in fire of 35 min duration	0.41	The probability of exceeding 35 min (relief valve lifts) is evaluated from Fig. 3-10 as $1 - 0.59 = 0.41$.

TABLE A-2. ***Continued***

Parameter	Probability of failure	Basis for evaluation
Weakened or defective tank with insulation damage fails in fire of 35 min duration	0.41	The probability of exceeding 35 min (relief valve lifts) is evaluated from Fig. 3-10 as $1 - 0.59 = 0.41$.
Tank with insulation damage ejects $> 50\%$ of inventory owing to fire of 100 min	0.12	The probability of exceeding 100 min is evaluated from Fig. 3-10 as $1 - 0.88 = 0.12$.
Relief valve of tank with insulation damage lifts from fire, but ejection of $> 50\%$ of inventory does not occur	0.29	The probability of a fire exceeding 35 min (relief valve lifts) but not exceeding 100 min ($> 50\%$ inventory loss) is $0.41 - 0.12 = 0.29$.
Tank with insulation damage and liquid discharge fails in fire due to overpressure, 55 min	0.18	The probability of exceeding 55 min is evaluated from Fig. 3-10 as $1 - 0.82 = 0.18$.
Relief valve of tank with insulation damage lifts from fire, but overpressurization does not occur with only liquid discharge	0.23	The probability of a fire exceeding 35 min (relief valve lifts) but not exceeding 55 min (overpressurization) is $0.41 - 0.18 = 0.23$.

are sufficient for the analyst to evaluate the fault tree in Figs. 4-8 and 4-9 and to complete the numerical entries for "Train" in Table 8-2.

FRV—Relief valve lifts, tank upright, vapor release
Fire occurs $= 5.21 \times 10^{-5}$ fires/180 tons transported
Fire duration between 100 and 290 min $= 0.11$ failures/fire
Tank car is upright $= 0.69$ upright failures/failure
FRV frequency $= 4.0 \times 10^{-6}$ failures/180 tons transported

FRL—Relief valve lifts, tank overturned, liquid release
Fire occurs $= 5.21 \times 10^{-5}$ fires/180 tons transported
Fire duration between 100 and 164 min $= 0.06$ failures/fire
Tank car is overturned $= 0.31$ overturned tank/event
FRL frequency $= 9.7 \times 10^{-7}$ failures/180 tons delivered

FFF—Relief valve fails to lift, tank fails
Fire occurs $= 5.21 \times 10^{-5}$ fires/180 tons delivered
Fire duration exceeds 100 min $= 0.12$ failures/fire
Relief valve fails to open $= 2.1 \times 10^{-4}$ valve failures/demand
FFF frequency $= 1.3 \times 10^{-9}$ failures/180 tons delivered

FFL—Relief valve lifts, tank car overturned, tank fails
Fire occurs = 5.21×10^{-5} fires/180 tons delivered
Fire duration exceeds 164 min = 0.06 failures/fire
Tank car is overturned = 0.31 overturned tank/event
FFL frequency = 9.7×10^{-7} failures/180 tons delivered

FFI—Relief valve lifts, tank fails when 50% inventory loss occurs
Fire occurs = 5.21×10^{-5} fires/180 tons delivered
Fire duration exceeds 290 min = 0.01 failures/fire
Tank is upright = 0.69 upright tank/event
FFI frequency = 3.6×10^{-7} failures/180 tons delivered

FD—Fire fails defective tank
Fire occurs = 5.21×10^{-5} fires/180 tons delivered
Fire duration exceeds 100 min = 0.12 failures/fire
Tank is defective = 2.6×10^{-3} defective tank/event
FD frequency = 1.6×10^{-8} failures/180 tons delivered

FW—Fire fails weakened tank. As defined in the text for this example, all impacts that weaken the tank also cause insulation damage. The FWI scenario includes the FW scenario.

FWI—Weakened tank with insulation damage fails in fire
Impact accident occurs = 2.32×10^{-3} impacts/180 tons delivered
Impact weakens tank = 1.50×10^{-2} weakened tank/event
Fire occurs after impact = 1.0×10^{-2} fires/impact
Fire duration exceeds 35 min = 0.41 failures/fire
FWI frequency = 1.4×10^{-7} failures/180 tons delivered

FRVI—Insulation damage, relief valve lifts, tank upright, vapor release
Impact accident occurs = 2.32×10^{-3} impacts/180 tons delivered
Impact damages only insulation = 9.18×10^{-2} damaged insulation/event
Fire occurs after impact = 1.0×10^{-2} fires/impact
Fire duration between 35 and 100 min = 0.29 failures/fire
Tank car is upright = 0.69 upright tank/event
FRVI frequency = 4.3×10^{-7} failures/180 tons delivered

FRLI—Insulation damaged, relief valve lifts, tank overturned, liquid release
Impact accident occurs = 2.32×10^{-3} impacts/180 tons delivered
Impact damages only insulation = 9.18×10^{-2} damaged insulation/event
Fire occurs after impact = 1.0×10^{-2} fires/impact
Fire duration between 35 and 55 min = 0.23 failures/fire
Tank car is overturned = 0.31 overturned tank/event
FRLI frequency = 1.5×10^{-7} failures/180 tons delivered

FFFI—Insulation damaged, relief valve fails, tank fails
Impact accident occurs = 2.32×10^{-3} impacts/180 tons delivered
Impact damages only insulation = 9.18×10^{-2} damaged insulation/event
Fire occurs after impact = 1.0×10^{-2} fires/impact
Fire duration exceeds 35 min = 0.41 failures/fire
Relief valve fails = 2.1×10^{-4} valve failures/demand
FFFI frequency = 1.8×10^{-10} failures/180 tons delivered

FFLI—Insulation damaged, relief valve lifts, tank overturned, tank fails
Impact accident occurs = 2.32×10^{-3} impacts/180 tons delivered
Impact damages only insulation = 9.18×10^{-2} damaged insulation/event
Fire occurs after impact = 1.0×10^{-2} fires/impact
Fire duration exceeds 55 min = 0.18 failures/fire
Tank car overturned = 0.31 overturned tank/event
FFLI frequency = 1.2×10^{-7} failures/180 tons delivered

FFII—Insulation damaged, relief valve lifts, tank fails when >50% inventory loss occurs
Impact accident occurs = 2.32×10^{-3} impacts/180 tons delivered
Impact damages only insulation = 9.18×10^{-2} damaged insulation/event
Fire occurs after impact = 1.0×10^{-2} fires/impact
Fire duration exceeds 100 min = 0.12 failures/fire
Tank car upright = 0.69 upright tank/event
FFII frequency = 1.8×10^{-7} failures/180 tons delivered

A.2 TRUCK ACCIDENT SCENARIOS

Truck Impact Accidents

The basic event "impact occurs on tank head" in Fig. 4-5 includes three factors: the frequency that an accident occurs, the conditional probability that an impact force results from the accident, and the conditional probability that the impact force is applied to the tank head. The accident frequency for the State Route 230 option was determined in Section 8.4.1 to be 3.94×10^{-3} accidents/180 tons delivered. Impact forces arise from collision accidents, which constitute 80% of all accidents. The velocity vectors used to determine Table 3-12 were based on 86% along the truck axis and 14% perpendicular to the truck axis. The basic event "impact occurs on tank head" can now be evaluated as:

(3.94 × 10^{-3} truck accidents/180 tons)
× (0.80 impact accidents/truck accident)
× (0.86 head impacts/impact accidents = 2.71×10^{-3} head impacts/180 tons

The conditional probability of impact on the manway or the valve enclosure, given shell impact, is the same as for a train, that is, 1.18×10^{-1}. The conditional probability of impact on the manway or the valve enclosure, given any impact, is 1.65×10^{-2}. The frequency of a manway or valve cover impact is:

$$
\begin{aligned}
&(3.94 \times 10^{-3} \text{ truck accidents/180 tons}) \\
&\times (0.80 \text{ impact accidents/truck accident}) \\
&\times (1.65 \times 10^{-2} \text{ manway impacts/impact accident}) \\
&= 5.20 \times 10^{-5} \text{ manway impacts/180 tons}
\end{aligned}
$$

The frequency of shell impact is evaluated similarly as:

$$
\begin{aligned}
&(3.94 \times 10^{-3} \text{ truck accidents/180 tons}) \\
&\times (0.80 \text{ impact accidents/truck accident}) \\
&\times (0.24 \text{ shell impacts/impact accident}) = 3.90 \times 10^{-4} \text{ shell impacts/180 tons}
\end{aligned}
$$

The probability that the impact force magnitude is large enough to cause failure that leads to a release is evaluated from Table 3-12. The conditional probabilities of impact failure, given an impact, are given in Table A-3. The transport vehicle weight is 25 tons. The impact scenarios now can be evaluated.

ILV—Impact fails liquid valve and excess flow valve does not close
Impact occurs on valve dome = 5.20×10^{-5} valve impacts/180 tons transported
Impact sufficient to fail liquid valve = 2.19×10^{-1} failures/impact
Excess flow valve fails to close = 2.2×10^{-3} failures/demand
ILV frequency = 2.5×10^{-8} failures/180 tons transported

IGV—Impact fails safety valve (tank trucks have excess flow valves on gas valve)
Impact occurs on valve dome = 5.20×10^{-5} valve impacts/180 tons transported
Impact sufficient to fail gas valves = 2.19×10^{-1} failures/impact
IGV frequency = 1.1×10^{-5} failures/180 tons transported

ISD—Impact fails defective tank shell
Impact occurs on shell = 3.90×10^{-4} shell impacts/180 tons transported
Shell is defective = 2.6×10^{-3} defective shell/shell
Impact sufficient to fail defective shell = 4.39×10^{-2} failures/impact
ISD frequency = 4.4×10^{-8} failures/180 tons transported

TABLE A-3. Probabilities of truck example impact failures

Parameter	Probability given an impact accident	Basis for evaluation
Impact fails tank head	9.0×10^{-4}	The head fails at 56 mph from impact (Table 5-2). The probability of exceeding this impact velocity is $1 - 0.9991 = 0.0009$ per impact accident.
Impact fails tank shell	9.4×10^{-3}	The shell fails at 32 mph from impact (Table 5-2). The probability of exceeding this impact velocity is $1 - 0.9906 = 0.0094$ per impact accident.
Impact fails liquid or gas valves	2.19×10^{-1}	Given that impact occurs on the valve dome, a 5-mph velocity is assumed to fail the valves. The probability of exceeding 5 mph is $1 - 0.7813 = 0.2187$.
Impact fails manway	9.4×10^{-3}	The failure of interest is the connection between the tank shell and the manway cover. The 32-mph value for the shell is used as the failure threshold.
Impact fails defective head	4.39×10^{-2}	The failure threshold is assumed to be 16 mph. The probability of exceeding this impact velocity is $1 - 0.9561 = 0.0439$.
Impact fails defective shell	4.39×10^{-2}	The failure threshold is assumed to be 16 mph.
Impact fails defective manway	4.39×10^{-2}	The failure threshold is assumed to be the same as that for a defective shell.

IS—Impact fails tank shell

Impact occurs on shell = 3.90×10^{-4} shell impacts/180 tons transported
Impact sufficient to fail shell = 9.4×10^{-3} failures/impact
IS frequency = 3.7×10^{-6} failures/180 tons transported

IHD—Impact fails defective tank head

Impact occurs on head = 2.71×10^{-3} head impacts/180 tons transported
Head is defective = 2.6×10^{-3} defective head/head
Impact sufficient to fail defective head = 4.39×10^{-2} failures/impact
IHD frequency = 3.1×10^{-7} failures/180 tons transported

IH—Impact fails tank head

Impact occurs on head = 2.71×10^{-3} head impacts/180 tons transported
Impact sufficient to fail head = 9.0×10^{-4} failures/impact
IH frequency = 2.4×10^{-6} failures/180 tons transported

IMD—Impact fails defective manway
Impact occurs on manway = 5.20×10^{-5} manway impacts/180 tons transported
Manway is defective = 2.6×10^{-3} defective manway/manway
Impact sufficient to fail defective manway = 4.39×10^{-2} failures/impact
IMD frequency = 5.9×10^{-9} failures/180 tons transported

IM—Impact fails manway
Impact occurs on manway = 5.20×10^{-5} manway impacts/180 tons transported
Impact sufficient to fail manway = 9.4×10^{-3} failures/impact
IM frequency = 4.9×10^{-7} failures/180 tons transported

The probability that a truck accident producing impact forces, that is, a collision accident, also produces a fire is 0.024 (Section 3.2.1.1). The frequencies of the impact scenarios and the impact followed by fire scenarios now can be entered into Table 8-2.

Truck Puncture Accidents

The tank shell and head thickness is 0.625 in. (Table 5-2). The probability of a puncture for this thickness, obtained (by extrapolation) from Table 3-13, is 1.69×10^{-3} punctures/accident. The valve dome thickness is 0.375 in., but the minimum thickness value in Table 3-13 of 0.4375 in. is used, and the result is 1.77×10^{-3} failures/accident. These units mean that the conditional probability that a puncture situation occurs, given an accident, has been included in the values obtained from Table 3-13.

Impact and puncture forces both arise from collision accidents; therefore, the distribution derived for head/shell impact location is assumed to be applicable for puncture location. The conditional probability of puncture on the valve dome is estimated as 0.003 on the basis of the area ratio of the dome to the shell. The head/shell/dome probabilities of puncture location are 0.86/0.139/0.001. The dome puncture location probability has been subtracted from the value for the shell puncture location probability.

If a probe punctures the valve dome, then on the basis of the number of valves present, it is assumed to fail a liquid valve 40% of the time, a gas valve 40% of the time, and a safety valve 20% of the time. The assumption that each valve dome puncture will result in the failure of one valve will tend to offset the lower computed valve cover penetration probability due to the limitations of Table 3-13. The probability of multiple valve punctures is assumed to be negligibly small.

The use of values from Table 3-13 results in the units for the "probe impact sufficient to penetrate head" event that includes aspects of the "puncture situations occurs" event, as the two are defined in Fig. 4-6. As was the case for the train puncture evaluation, the use of explicit units keeps the calculations correct, regardless of how the conflict between definitions is resolved.

PH—Puncture probe fails tank head
Truck accident occurs = 3.94×10^{-3} accidents/180 tons delivered
Puncture probe contacts head = 0.86 head punctures/puncture
Puncture failure occurs = 1.69×10^{-3} punctures/accident
PH frequency = 5.7×10^{-6} punctures/180 tons delivered

PS—Puncture probe fails tank shell
Truck accident occurs = 3.94×10^{-3} accidents/180 tons delivered
Puncture probe contacts shell = 0.139 shell punctures/puncture
Puncture failure occurs = 1.69×10^{-3} punctures/accident
PS frequency = 9.2×10^{-7} punctures/180 tons

PV—Puncture probe fails valve dome
Truck accident occurs = 3.94×10^{-3} accidents/180 tons delivered
Puncture probe contacts dome = 0.001 dome punctures/puncture
Puncture failure occurs = 1.77×10^{-3} punctures/accident
PV frequency = 7.0×10^{-9} punctures/180 tons

The probability that a truck accident producing puncture forces, that is, a collision accident, also produces a fire is 0.024 (Table 3-10). The frequencies of the puncture scenarios and the puncture followed by fire scenarios now can be entered into Table 8-2.

Truck Crush Accidents

The crush threshold from Table 5-2 is about twice the maximum force of 2000 lb/ft (Section 3.2.1.3). Therefore, there is no credible crush failure; crush failure frequencies are zero.

Truck Fire Accidents

The probability of fire, given a truck accident, is 0.016 (Section 3.2.1.1). The "fire occurs" basic event is then:

$$(3.94 \times 10^{-3} \text{ accidents/180 tons delivered})(1.6 \times 10^{-2} \text{ fires/accident}) = 6.30 \times 10^{-5} \text{ truck fires/180 tons}$$

Impacts between the value required to fail a defective shell (16 mph) and the value required to fail a nondefective shell (32 mph) are assumed to substantially weaken the shell and cause a 10% insulation loss. The conditional probability of a weakened container with insulation damage, given an impact accident, is the difference between the 16- and 32-mph values in Table A-3, or 3.45×10^{-2}. Insulation damage but no degradation in tank strength is assumed to occur for impact accidents less than 16 mph but greater than 5 mph. The conditional probability of insulation damage only, given an impact accident, is the difference between the 5- and 16-mph values from Table A-3, or 1.75×10^{-1}. The conditional probability of a fire following impact is 0.024 (Table 3-10).

The probability of a truck being upright in an impact accident is estimated as the fraction of collisions along the truck axis, 0.86 (Dennis et al. 1978). The probability that a truck involved in a fire is upright is estimated from Table 3-10. Upright trucks are assumed to occur in all of the noncollision, fire-only accidents and in the fraction of the collision fires that are collisions along the truck axis; this value is 0.895.

The conditional probabilities of truck fire failures for this example, using the fire failure thresholds in Table 5-2, are shown in Table A-4. The scenario frequencies now are evaluated.

FRV—Relief valve lifts, tank upright, vapor release
Fire occurs $= 6.3 \times 10^{-5}$ truck fires/180 tons
Fire duration between 50 and 145 min $= 1.9 \times 10^{-2}$ failures/fire
Tank truck is upright $= 0.895$ upright failures/failure
FRV frequency $= 1.1 \times 10^{-6}$ failures/180 tons delivered

TABLE A-4. Probabilities of truck example fire failures

Parameter	Probability of failure	Basis for evaluation
Fire duration raises pressure to relief valve pressure, 50 min	2.0×10^{-2}	The probability of exceeding 50 min is evaluated from Fig. 3-2 as $1 - 0.98 = 0.02$.
Fire duration fails weakened or defective shell, 50 min	2.0×10^{-2}	The container is assumed to rupture at the relief valve start-to-discharge pressure of 1.65×10^{6} N/m^2 (226 psig).
Fire duration ejects > 50% of inventory, 145 min	1.1×10^{-3}	The probability of exceeding 145 min is evaluated from the equation in Fig. 3-2 as 1.1×10^{-3}.
Fire duration lifts relief valve, but ejection of > 50% inventory does not occur	1.9×10^{-2}	The probability of a fire that exceeds 50 min (relief valve lifts) but does not exceed 145 min (> 50% loss) is $2.0 \times 10^{-2} - 1.1 \times 10^{-3} = 1.9 \times 10^{-2}$.
Weakened or defective shell with insulation damage fails in fire of 18 min duration	1.7×10^{-1}	The container is assumed to rupture at the relief valve start-to-discharge pressure of 1.65×10^{-6} N/m^2 (225 psig).

TABLE A-4. ***Continued***

Parameter	Probability of failure	Basis for evaluation
Tank with insulation damage ejects $>50\%$ of inventory owing to fire of 50 min	2.0×10^{-2}	The probability of exceeding 50 min is $1 - 0.98 = 0.02$.
Relief valve of tank with insulation damage lifts from fire, but ejection of $>50\%$ inventory does not occur	1.5×10^{-2}	The probability of a fire that exceeds 18 min (lifts relief valve) but does not exeed 50 min ($>50\%$ loss) is $1.7 \times 10^{-1} - 2.0 \times 10^{-2} = 1.5 \times 10^{-2}$.
Tank with insulation damage and liquid discharge fails in fire owing to overpressure, 28 min	6.0×10^{-2}	The probability of exceeding 28 min is $1 - 0.94 = 0.06$.
Relief valve of tank with insulation damage lifts owing to fire, but overpressurization does not occur with liquid discharge	1.1×10^{-1}	The probability of a fire that exceeds 18 min (lifts relief valve) but does not exceed 28 min (overpressurization) is $1.7 \times 10^{-1} - 6.0 \times 10^{-2} = 1.1 \times 10^{-1}$.
Fire duration causes overpressure with liquid discharge, 82 min	3.2×10^{-3}	The probability of exceeding 82 min is evaluated from the equation in Fig. 3-2 as 3.2×10^{-3}.
Fire duration lifts relief valve, but there is not overpressure with liquid discharge	1.7×10^{-2}	The probability of a fire that exceeds 50 min (relief valve lifts) but does not exceed 82 min (overpressurization) is $2.0 \times 10^{-2} - 3.2 \times 10^{-3} = 1.7 \times 10^{-2}$.
Relief valve of tank with insulation damage lifts in fire of 18 min duration	1.7×10^{-1}	The probability of exceeding 18 min is $1 - 0.83 = 0.17$.
Weakened or defective shell with insulation damage fails in fire of 18 min duration	1.7×10^{-1}	The container is assumed to rupture at the relief valve start-to-discharge pressure of 1.65×10^{-6} N/m^2 (225 psig).
Tank with insulation damage ejects $>50\%$ of inventory owing to fire of 50 min	2.0×10^{-2}	The probability of exceeding 50 min is $1 - 0.98 = 0.02$.
Relief valve of tank with insulation damage lifts from fire, but ejection of $>50\%$ inventory does not occur	1.5×10^{-2}	The probability of a fire that exceeds 18 min (lifts relief valve) but does not exceed 50 min ($>50\%$ loss) is $1.7 \times 10^{-1} - 2.0 \times 10^{-2} = 1.5 \times 10^{-2}$.
Tank with insulation damage and liquid discharge fails in fire owing to overpressure, 28 min	6.0×10^{-2}	The probability of exceeding 28 min is $1 - 0.94 = 0.06$.
Relief valve of tank with insulation damage lifts owing to fire, but overpressurization does not occur with liquid discharge	1.1×10^{-1}	The probability of a fire that exceeds 18 min (lifts relief valve) but does not exceed 28 min (overpressurization) is $1.7 \times 10^{-1} - 6.0 \times 10^{-2} = 1.1 \times 10^{-1}$.

FRL—Relief valve lifts, tank overturned, liquid release
Fire occurs = 6.3×10^{-5} truck fires/180 tons
Fire duration between 50 and 82 min = 1.7×10^{-2} failures/fire
Tank truck is overturned = 0.105 overturn failures/failure
FRL frequency = 1.1×10^{-7} failures/180 tons delivered

FFF—Relief valve fails to lift, tank fails
Fire occurs = 6.3×10^{-5} truck fires/180 tons
Fire duration exceeds 50 min = 2.0×10^{-2} failures/fire
Relief valve fails to open = 2.1×10^{-4} valve failures/demand
FFF frequency = 2.6×10^{-10} failures/180 tons delivered

FFL—Relief valve lifts, truck overturned, tank fails
Fire occurs = 6.3×10^{-5} truck fires/180 tons
Fire duration exceeds 82 min = 3.2×10^{-3} failures/fire
Tank truck is overturned = 0.105 overturn failures/failure
FFL frequency = 2.1×10^{-8} failures/180 tons delivered

FFI—Relief valve lifts, tank fails when 50% inventory loss occurs
Fire occurs = 6.3×10^{-5} truck fires/180 tons
Fire duration exceeds 145 min = 1.1×10^{-3} failures/fire
Tank truck is upright = 0.895 upright failures/failure
FFI frequency = 6.2×10^{-8} failures/180 tons delivered

FD—Fire fails defective tank
Fire occurs = 6.3×10^{-5} truck fires/180 tons
Fire duration exceeds 50 min = 2.0×10^{-2} failures/fire
Tank is defective = 2.6×10^{-3} defective tank/event
FD frequency = 3.3×10^{-9} failures/180 tons delivered

FW—Fire fails weakened tank. As defined in the text for this example, all impacts that weaken the tank also cause insulation damage. The FWI scenario includes the FW scenario.

FWI—Weakened tank with insulation damage fails in fire
Impact accident occurs = 3.15×10^{-3} impacts/180 tons delivered
Impact weakens tank = 3.45×10^{-2} weakened tank/event
Fire occurs after impact = 2.4×10^{-2} fires/impact
Fire exceeds 18 min = 1.7×10^{-1} failures/fire
FWI frequency = 4.4×10^{-7} failures/180 tons delivered

FRVI—Insulation damage, relief valve lifts, tank upright, vapor release
Impact accident occurs = 2.32×10^{-3} impacts/180 tons delivered
Impact damages only insulation = 9.18×10^{-2} damaged insulation/event
Fire occurs after impact = 1.0×10^{-2} fires/impact
Fire duration between 35 and 100 min = 0.29 failures/fire
Tank car is upright = 0.69 upright tank/event
FRVI frequency = 4.3×10^{-7} failures/180 tons delivered

FRLI—Insulation damaged, relief valve lifts, tank overturned, liquid release
Impact accident occurs = 2.32×10^{-3} impacts/180 tons delivered
Impact damages only insulation = 9.18×10^{-2} damaged insulation/event
Fire occurs after impact = 1.0×10^{-2} fires/impact
Fire duration between 35 and 55 min = 0.23 failures/fire
Tank car is overturned = 0.31 overturned tank/event
FRLI frequency = 1.5×10^{-7} failures/180 tons delivered

FFFI—Insulation damaged, relief valve fails, tank fails
Impact accident occurs = 2.32×10^{-3} impacts/180 tons delivered
Impact damages only insulation = 9.18×10^{-2} damaged insulation/event
Fire occurs after impact = 1.0×10^{-2} fires/impact
Fire duration exceeds 35 min = 0.41 failures/fire
Relief valve fails = 2.1×10^{-4} valve failures/demand
FFFI frequency = 1.8×10^{-10} failures/180 tons delivered

FFLI—Insulation damaged, relief valve lifts, tank overturned, tank fails
Impact accident occurs = 2.32×10^{-3} impacts/180 tons delivered
Impact damages only insulation = 9.18×10^{-2} damaged insulation/event
Fire occurs after impact = 1.0×10^{-2} fires/impact
Fire duration exceeds 55 min = 0.18 failures/fire
Tank car overturned = 0.31 overturned tank/event
FFLI frequency = 1.2×10^{-7} failures/180 tons delivered

FFII—Insulation damaged, relief valve lifts, tank fails when >50% inventory loss occurs
Impact accident occurs = 2.32×10^{-3} impacts/180 tons delivered
Impact damages only insulation = 9.18×10^{-2} damaged insulation/event
Fire occurs after impact = 1.0×10^{-2} fires/impact
Fire duration exceeds 100 min = 0.12 failures/fire
Tank car upright = 0.69 upright tank/event
FFII frequency = 1.8×10^{-7} failures/180 tons delivered

References

Note: The reports of U.S. government agencies, their laboratories, and contractors cited here are available from the National Technical Information Service, Springfield, Virginia 22161, USA.

Andrews, W. B., et al. March 1980. *An Assessment of the Risks of Transporting Liquid Chlorine by Rail.* PNL-3376. Pacific Northwest Laboratory.

CCPS (Center for Chemical Process Safety). 1989. *Guidelines for Process Equipment Reliability Data with Data Tables.* New York: American Institute of Chemical Engineers.

Dennis, A. W., et al. 1978. *Severities of Transportation Accidents Involving Large Packages.* SAND77-0001. Sandia National Laboratories.

Southwest Research Institute. September 1979. *Nuclear Plant Reliability Data System, 1978 Annual Report of Cumulative System and Component Reliability.* NUREG/CR-0942. U.S. Nuclear Regulatory Commission.

Appendix B

Characterization and Aggregation of Source Terms

The release scenarios of the accident scenarios listed in Table 8-2 are characterized in this appendix. To keep evaluation costs reasonable, scenarios with similar source terms are grouped. The results of this analysis are given in Table 8-5.

B.1 TRAIN SOURCE TERM INITIAL CHARACTERIZATION

ILV—Impact Fails Liquid Valve and Excess Flow Valve Does Not Close

This scenario also can occur from the valve dome puncture scenario (PV) if the liquid valve is struck. The vapor and liquid phases are at ambient temperature at the corresponding vapor pressure plus any added padding. For chlorine at 25°C (77°F), the vapor pressure is about 97 psig (The Chlorine Institute 1986). The liquid valves are connected to eduction tubes that extend to the bottom of the tank; thus, the effective hole in the tank car is 180 degrees from the apparent hole. If the tank car is nearly upright, the initial flow is from the liquid space, and a portion of the superheated liquid will flash to vapor (Eq. 6-6). If the tank car is upside down, the initial flow is from the vapor space. The flow may start from either the vapor or the liquid space and then change to the other space as the liquid level drops. As the pressure drops, liquid chlorine will flash to vapor inside the tank car, reducing the rate of pressure drop. The discharge flow rate will decrease as the pressure drops. About 19% of the liquid will flash as the pressure drops

(Eq. 6-6). Depending on the tank orientation, boiling liquid chlorine could continue to flow out of the tank owing largely to the gravity head. When boiling liquid no longer drains, chlorine vapor, produced by evaporation, will be released. Three discrete cases can be used to approximate the various situations that can occur:

1. The tank car is nearly upright. The initial discharge is superheated liquid driven by the high tank pressure. After the pressure drops to the ambient pressure, subsequent liquid release occurs as pressure builds up sufficiently to overcome the gravity head in the eduction tube.
2. The tank car is overturned approximately 90 degrees. The initial liquid discharge is followed by draining of the boiling liquid. After 50% total release, further release occurs only as liquid evaporates.
3. The tank car is nearly completely overturned, that is, 180 degrees. The initial release is from the vapor space followed by a smaller evaporation-produced release.

IGV—Impact Fails Gas Valve

This scenario also occurs for the valve dome puncture scenario, PV, if the gas valve is struck. The possible source terms are similar to those for the ILV scenario except that, without eduction tubes, 100% drainage now can occur from the gas valve, depending on the tank car orientation. Three discrete cases can be used to approximate the various situations that can occur:

1. The tank car is nearly upright. The initial vapor discharge is followed by a smaller evaporation-produced release of vapor.
2. The tank car is overturned approximately 90 degrees. The initial liquid discharge is followed by draining of the boiling liquid. After 50% total release, release occurs as liquid evaporates.
3. The tank car is nearly completely overturned. The initial release is from the liquid space followed by essentially complete drainage of the boiling liquid.

RI—Nonaccident Release from Leaking Valve

This nonaccident release will be from an upright tank car. If the release is due to the safety valve functioning accidentally, the flow rate will be the same as for the IGV scenario for an upright tank car. A loose valve would present a much smaller flow area than that for a functioning valve and a corresponding smaller release rate.

ILVF, IGVF, and PVF—Impact or Puncture Fails Either Gas or Liquid Valve Followed by Fire

The release scenarios are similar to the three valve scenarios described previously. The fire will significantly increase the rate of evaporation inside and outside the tank car. The effective temperature of all releases will be increased, thereby increasing the downwind dispersion.

IS, ISD, IH, IHD, IM, IMD, PH, PS, CD, C—Impact, Puncture, and Crush of the Tank Car

A spectrum of hole sizes can result from impact, puncture, and crush accident scenarios. Each of the three types of accident is likely to have a unique spectrum of hole sizes and locations. As data are generally lacking to characterize them, these releases are aggregated into four discrete cases:

1. A small hole occurs slightly above the tank centerline. The initial superheated liquid discharge will be followed by draining of the boiling liquid. After about 50% total release, release occurs as the liquid evaporates.
2. A small hole occurs near the bottom of the tank. The initial superheated liquid discharge will be followed by essentially complete drainage of the boiling liquid.
3. A large hole occurs slightly above the tank centerline. The discharge rates are larger than for the analogous small-hole case.
4. A large hole occurs near the bottom of the tank. The discharge rates are larger than for the analogous small-hole case.

ISF, ISDF, IHF, IHDF, IMF, IMDF, PHF, PSF, CDF, CF—Impact, Puncture, and Crush Failure Followed by Fire

The release scenarios are similar to the corresponding scenarios without fire already described. The fire will increase evaporation rates and increase downwind dispersion.

FW, FWI, FD, FFF, FFFI—Rapid Release of Entire Contents at Elevated Temperatures

These scenarios all involve the sudden failure of the tank car because of high pressure at high temperature. The contents are released rapidly into a fire environment that further increases the average temperature of the tank car contents.

FFI, FFII, FFL, FFLI—Rapid Release of a Large Amount, Preceded by Vapor or Superheated Liquid Release

In these four scenarios, vapor or superheated liquid is released at above-ambient temperatures into a fire environment for 20 to 190 min. Then the tank car fails suddenly, and it releases on the order of 50% of the original tank contents into a fire environment.

FRV, FRVI, FRL, FRLI—Fire Lifts Relief Valve

In these four scenarios, the relief valve is lifted owing to high tank pressures caused by a fire. The fire terminates in 20 to 190 min, and the relief valve closes to terminate the release. Depending on the tank orientation, the release is from either the vapor or the liquid space.

Summary

The source term characterization described in this section shows how similar scenarios can be aggregated to reduce the number of scenarios. At the same time, multiple characteristics of the releases produced by the scenarios are identified, and this increases the number of scenarios. The net result is a reduction of the 40 original accident scenarios to 23. Each scenario applies to both truck and train; therefore, a total of 46 scenarios must be evaluated. In the next section, these scenarios are aggregated further to reduce the cost of consequence evaluation.

B.2 TRAIN SOURCE TERM FINAL CHARACTERIZATION

The number of consequence calculations is reduced to 23 in Section B.1, but the effort needed to evaluate the consequences of this number of scenarios is considered excessive for this example, as for many practical decisions. In this section, the number of consequences is reduced by further aggregation of scenarios.

Mechanical Damage to Valves

The ILV, IGV, ILVF, IGVF, PV, and PVF scenarios are similar in that the initial large release rate of either superheated vapor or superheated liquid is followed by a longer period with a small release rate. Of these six scenarios, the frequency of the IGV scenario is two or more orders of magnitude larger than the other frequencies (Table 8-2). Therefore, the IGV scenario is selected to represent this class of release. Three potential tank car orientations were identified in Section B.1 to determine the release characterization of the IGV scenario. The conditional probability of impact on the valve dome evaluated in Appendix A is largely based on a tank car orientation of approximately 90 degrees; therefore, this orientation is selected to represent the IGV scenario. In this orientation, the initial release is from the liquid space. Some of the released superheated liquid will flash, and the remaining liquid will be completely aerosolized (Section 6.2). The temperature of the vapor and aerosol will be the boiling point, 239°K. The source term is described in Table 8-5. A restriction for many dispersion computer programs is that the discharge rate must be constant; therefore, many analysts conservatively use the initial discharge rate as that constant value. The consequence analyses in this book use the same approach. The initial discharge of superheated liquid will dominate the later, smaller release rate.

Mechanical Damage to Head, Shell, or Manway

The remaining mechanical failure and mechanical failure followed by fire scenarios are similar to the valve mechanical failure in that the initial large release (generally of superheated liquid) is followed by a longer period of time characterized by a smaller release rate. Releases from damage to the relief and liquid valves discussed previously are limited to holes no larger than 1 in. The mechanical failures addressed in this set of scenarios are expected to produce holes in the tank that are as small as about 1 in. diameter and may be larger. The IS, impact damage to shell, scenario is the most probable and will be used to represent this class of release. The other scenarios represented by the IS scenario are ISF, ISD, ISDF, IHD, IHDF, IH, IHF, IMD, IMDF, IM, IMF, CD, CDF, C, CF, PH, PHF, PS, and PSF. Two sizes are used in Section B.1 to characterize the spectrum of holes that could be produced. In more elaborate analyses, three or more hole sizes are used to represent the range of possibilities. One hole size, 20 cm (8 in.), is used in this example. The initial liquid discharge rate of 830 kg/s (1830 lb/s) is assumed to remain constant until the tank is empty. This assumption implies that the hole is near the bottom of the tank (case 4 in Section B.1).

Nonaccident Release

Of the 10 leaking chlorine tank cars in 1990 and 1991, one involved a functioned safety valve, six a loose safety valve, two a loose liquid valve plug, and one a loose manway nozzle (BOE 1991,1992). The functioned safety valve will produce the largest release rate owing to the larger flow area. The other releases are so much smaller that fatalities are not possible except in very unusual circumstances. It is then appropriate to reduce the overall scenario frequency, by an order of magnitude in this case. The source term specifications for the RI scenario are given in Table 8-5.

Relief Valve Lifted by Fire

The FRV, FRL, FRVI, and FRLI scenarios are similar in that the relief valve is actuated when the pressure in the tank reaches 2.69×10^6 N/m^2 (375 psig). The associated chlorine temperature is 82°C (180°F). The valve closes when the pressure drops after the fire stops. The FRV scenario is four times more likely than the FRL scenario and two orders of magnitude more likely than either of the FRVI or FRLI scenarios; therefore, it is selected to represent this release category. The four scenarios in this release category involve four release times, three different relief rates, and two different total releases. The FRV scenario has the smallest relief rate but the largest total release. Table 8-5 gives the discharge rate for this release category. The presence of the fire external to the tank car is conservatively assumed to have no effect on the escaping vapor. The car is upright 69% of the time in a fire (Section 3.2.3.1), the release is occurring 14.5 ft above the ground, and jet forces will be associated with the discharge that will propel the released vapor away from any fire. These factors tend to limit the influence of the fire on the discharge.

Rapid Release of Entire Contents at Elevated Temperature

The remaining fire scenarios involve a sudden release of all, or a substantial fraction, of the tank car contents. The scenarios are based on the tank car being immersed in a large fire; the most probable of these is the FWI scenario, "fire fails tank that has been weakened by impact." The FWI scenario will represent the following scenarios: FFF, FFL, FFI, FD, FW, FWI, FFFI, FFLI, and FFII. The impact also removes insulation from the tank. The fire is assumed to heat the released material from the release conditions at 2.69×10^6 N/m^2 (375 psig) of 82°C (180°F) to an average of 127°C (261°F). Average hydrocarbon fire temperatures are on the order of 980°C (1800°F); therefore, the effect of the fire on the released material is probably much greater than is assumed for this example. The effect of a thermal updraft in increasing dispersion from the fire that ruptured a chlorine car in the Mississauga, Canada, incident on October 11, 1979, is probably the reason why no one was killed even though 60 tons of chlorine was released (Marshall 1987). This type of release is characterized in Table 8-5.

B.3 TRUCK SOURCE TERM CHARACTERIZATION

The evaluation of train and truck scenario frequencies in Section 8.4.2 showed that descriptions of the scenarios are qualitatively the same in all but one case. The IGV truck scenario involves a safety valve rather than a gas valve because the truck gas valves have an excess flow valve. For all practical purposes, the discharge rate from an impact-damaged safety valve can be considered the same as the discharge rate from an impact-damaged gas valve. Therefore, there is no effective difference in the qualitative scenario descriptions. The five scenarios chosen in Section B.2 to characterize the consequences of train releases will be equally representative of truck release.

The gas, liquid, and relief valves are the same for the tank car as for the tank truck except that the truck gas valves also include an excess flow valve. Initial discharge rates at ambient conditions for valve releases, IGV and RI, will be the same for truck and train, as the hole size and the vapor pressure will be the same. The initial large hole discharge rate for the IS scenario will be the same in both cases for the same reason. In all three cases the duration of the discharge will be much shorter for truck releases compared with train releases because of the smaller truck inventory. The downwind plume travel time is shown to be short compared to the release time in either case; therefore, the dispersion is characterized by the release rate, not the total release, and the dispersion characteristics of each of these three releases will be the same for both truck and train.

The start-to-discharge relief valve setting for tank trucks is 225 psig (1.65×10^6 N/m^2) compared with 325 psig (2.34×10^6 N/m^2) for tank cars (The Chlorine Institute 1986). The FRV scenario discharge rate and release temperature for the two transport modes will reflect this difference, as shown in Table 8-5.

The FWI scenario is based on rupture and instantaneous release of the tank contents at the start-to-discharge pressure. The difference in the tank inventory and the temperature at the rupture pressure is reflected in Table 8-5. The effect of the fire on the temperature of the released chlorine is conservatively estimated to increase the average temperature by 45°C (113°F).

References

BOE (Bureau of Explosives). June 1991. *Report of Railroad Tank Car Leaks of Hazardous Materials by Commodity by Source of Leak for the Year 1990.* BOE 90-2. Washington, D.C.: Association of American Railroads.

BOE (Bureau of Explosives). June 1992. *Report of Railroad Tank Car Leaks of Hazardous Materials by Commodity by Source of Leak for the Year 1991.* BOE 91-2. Washington, D.C.: Association of American Railroads.

Marshall, V. C. 1987. *Major Chemical Hazards.* Chichester, England: Ellis Horwood Limited.

The Chlorine Institute. 1986. *The Chlorine Manual Fifth Edition.* Washington, D.C.: The Chlorine Institute, Inc.

Index